Mississippi R.

Rocky Mountains

SMITHSONIAN NATURAL HISTORY SERIES

John Kricher, Series Editor

Books in this series explore the diverse plants, animals, people, geology, and ecosystems of the world's most interesting environments, presented in an accessible style by world-renowned experts.

Rocky Mountains

Scott A. Elias

SMITHSONIAN
INSTITUTION
PRESS

WASHINGTON
& LONDON

For Sandy, who loved his patch of the Rockies

Copy editor: Debbie K. Hardin
Production editor: Robert A. Poarch
Designer: Brian Barth

Library of Congress Cataloging-in-Publication Data
Elias, Scott A.
Rocky Mountains / Scott A. Elias.
p. cm. — (Smithsonian natural history series)
Includes bibliographical references.
ISBN 1-58834-042-2 (alk. Paper)
1. Natural history—Rocky Mountains. I. Title. II. Series.
QH104.5.R6 E58 2002
508.78—dc21 2001057645

British Library Cataloguing-in-Publication Data is available

Manufactured in the United States of America
09 08 07 06 05 04 03 02 5 4 3 2 1

Color signature printed in China.

The recycled paper used in this publication meets the minimum requirements of the American National Standard for Information Sciences—Permanence of Paper for Printed Library Materials ANSI Z39.48-1984.

Contents

Editor's Note

One of the finest ecological treasures in North America is the vast "ribbon of rock" that runs through the continent, from Canada into Mexico: the Rocky Mountains. Seeing the ridge of snow-capped peaks for the first time, perhaps the Front Range that looms above as one approaches Denver from the east, is usually a moment to remember. The Rockies get your attention.

Late in the nineteenth century the American West was still being explored. It was then that C. Hart Merriam, head of the Division of Economic Ornithology and Mammalogy of the U.S. Department of Agriculture, organized an ambitious expedition to survey the complex ecology of the southern Rocky Mountains. He selected the San Francisco Peaks, north of Flagstaff, Arizona, and, on July 26, 1889, Merriam commenced two months of intensive field work documenting the various "life zones" encountered with changing elevation along the mountain slopes.

Merriam noted six life zones, from desert to mountaintop: Lower Sonoran, Upper Sonoran, Transition, Canadian, Hudsonian, and Alpine Tundra. The two sonoran zones were typically arid, supporting desert or small forests of pinyon pine and various juniper species. The Canadian and Hudsonian zones, as the name implies, were rich forests of spruce and fir, nearly identical in many respects to those found throughout the vast boreal forests of Canada. The Alpine Tundra was above treeline, ecologically similar to the most northern terrestrial ecosystem in the world, the arctic tundra. The Transition Zone was exactly that—an area where aridity gives way to increasing moisture and cooler temperatures, a land where forests of stately ponderosa pine (*Pinus ponderosa*) dominate.

It is a daunting task to capture the ecology of the Rocky Mountains in one book. After all, a trip from the grasslands of Ft. Collins to Trail Ridge Road in Rocky Mountain National Park is ecologically equivalent to making a latitudinal journey from Nebraska to Churchhill, Manitoba! And there is more.

The Rockies have a rich geologic history. Not far from where Merriam worked on the San Francisco Peaks, the Painted Desert and Petrified Forest speak to an age long gone, when small dinosaurs and lush semitropical forests covered landscape that is now desert. Where Utah and Colorado meet was once a vast floodplain where dinosaurs such as *Stegosaurus* and *Apatosaurus* tried to avoid the deadly jaws of *Allosaurus*. Known today as Dinosaur National Monument, the region has undergone vast changes in the 65 million years since the dinosaurs suffered the fate of extinction. Perhaps the geological gem of the lot, the Grand Canyon plunges deep as the surrounding mountains rise high, its depth revealing a 500-million-year-old tapestry of life forms preserved only as fossils.

I have authored a field guide to the ecology of Rocky Mountain forests, and so I have some appreciation of the challenge the region affords the would-be interpretive naturalist–author. It is indeed satisfying that Scott Elias, a recognized expert on Rocky Mountain geology and ecology, agreed to accept the task of writing a book for the Smithsonian Natural History series. Dr. Elias has plumbed the depths of time in his prolific research on paleoecological, paleoclimatic, zoogeographic, and evolutionary implications of insect fossil assemblages from the Quarternary period and has written on the ice age history of the Southwest. His knowledge of the region is exhaustive and his talent for sharing his understanding and enthusiasm for the unique Rocky Mountains will be more than evident in the pages that follow.

John Kricher

Preface

This book is about the natural history of a series of mountain ranges that divide a continent. They are the mountains that form one of the greatest geologic features of the Earth, the Rocky Mountains. We will explore their past and their present and consider their future as we examine the places that make the Rockies special, from northwestern Canada to the southwestern United States. We will converse about the plants and animals that live there today and those that remind us of their presence because of fossils left behind.

No place in North America evokes quite the sense of wonder and awe as the Rockies. They hold a special place in the hearts of millions—visitors and residents alike. There are many reasons that people are drawn here. Some seek the scenic splendor of lofty peaks, clear mountain lakes, glaciers, and waterfalls. Others come to enjoy the pleasures of mountain sports, from skiing and winter mountaineering to fly fishing and mountain biking. Many are attracted by the opportunity to view wildlife in natural habitats preserved in the many national parks of the Rockies. Some folks simply want to escape the summer heat of the lowlands. Of course none of these reasons for coming to the Rockies is mutually exclusive. A mountain biker may stop to admire a herd of elk in a meadow, and a bird watcher may enjoy sleeping in a warm sleeping bag as the nighttime temperatures drop near freezing, even in July. The Rockies evoke a sense of wonder because of the combination of these factors: their physical beauty, their unspoiled nature, their wildlife, and the challenges they offer to people engaged in a wide variety of mountain activities.

When you come to the Rockies, you sense that you have entered a place that is still wild and untamed, in spite of all that people have done to "civilize" the mountains. That sense of wilderness is becoming increasingly difficult to find in the world today. So much of the world is thoroughly tame, with no natural forests, no predators larger than a coyote or fox, and no lack of modern conveniences or transportation into every mountain valley and many mountaintops. But the Rockies are different. Here you can escape into true wilderness, just by walking a while on a mountain trail. Here you have the chance to see a bear, a mountain lion, or a wolf, just as Lewis and Clark saw them almost 200 years ago, when they first entered the Rocky Mountains on their "Voyage of Discovery." Here you can make your own personal discoveries, seeing things you have never seen before. That is what makes the Rockies exciting. I hope that you will let this book serve as a guide to this fascinating region, the backbone of the continent.

Scott A. Elias

Acknowledgments

This book was a true collaboration, and I must take a moment to thank my friends and colleagues for their help. My late father-in-law, Daniel Jobin, shared his knowledge of the bedrock geology and mineral resources. Paul Carrara of the U.S. Geological Survey in Denver provided advice on glacial geology. Mark Meier, Institute of Arctic and Alpine Research, University of Colorado (INSTAAR), gave advice on the section concerning snow. Mel Reasoner provided information on postglacial environments of the Canadian Rockies. William Bowman provided photos of alpine wildflowers from Niwot Ridge, Colorado. Mark Losleben provided Niwot Ridge climate data and gave advice on mountain climatology. Randle Robertson explained much about the Yoho-Burgess Shale. Darlene Koerner, Ashley National Forest, Utah, provided information concerning the Uinta Mountains. Yves Bosquet gave me information on the ground beetles of the Rockies. James Halfpenny of Naturalist's World provided many photos. Ralph Maughan discussed the reintroduction of wolves into Idaho and provided a map of Idaho wolf pack ranges. Finally, I thank my family for their forbearance as I spent many evenings and weekends working on this project.

Financial support for the preparation of this book was provided by a grant from the National Science Foundation, DEB-9211776 (Niwot Long Term Ecological Research, University of Colorado).

1

Backbone of a Continent

My awareness of the Rocky Mountains began at the tender age of four, when my father took me fishing in the southern Rockies west of Pueblo, Colorado. We fished all day, cooked hamburgers over a campfire, and slept in the back of the family station wagon (with a cavernous interior that was typical in American cars made in the 1950s). I was fascinated by that day spent on the edge of a clear, cool, cascading stream. It was such a wonder to me; the water ran and jumped and laughed its way over and around the rocks, boulders, and fallen trees in its path. In the backwaters the helter-skelter riffles settled into deep, silent pools, perfect for dreamy contemplation. We camped within earshot of the stream. Its rushing sounds lulled me to sleep that night, but not before I got my first, best look at the stars on a really clear night in the mountains. There was the Milky Way, something hardly ever seen from my backyard in Pueblo. The sheer *number* of stars was overwhelming. That night, my love for the Rocky Mountains was kindled.

The second turning point in my lifelong love affair with the Rockies took place a few years later, as a thirteen-year-old Boy Scout. Apart from picnics and fishing trips, my family never went camping. All that changed when I joined the Scouts. The Scouts taught me how to tie knots and memorize mottoes, but I was really in it for the camping. My first backpacking trip was a disaster. I was not in good shape, and my floppy, rucksack-style backpack lacked both padded shoulder straps and any kind of frame. The straps tormented my shoulders and the pack tugged at my back muscles. My new boots busied themselves gouging silver-dollar-sized blisters on my heels. My only goal was to survive the weekend, which I did. When the next backpacking trip came, I was better prepared. We hiked about 5 miles along a mountain trail, through aspen groves, along a mountain stream, and up to a huge rock formation at the head of a valley. We camped in a rock shelter formed where enormous sandstone boulders had weathered away at their bases. My body cooperated much more fully, so my brain was able to drink in the montane landscapes and enjoy the camaraderie of my camping buddies. I learned that the Rockies offer physical challenges and that meeting those challenges brings great satisfaction.

The third step in my Rocky Mountain odyssey began in my freshman year at the University of Colorado. I dove excitedly into plant and animal ecology courses and took many field trips into the Rockies near Boulder, memorizing the scientific names of plants and animals, insects, and lichens. Mountain ecology courses taught me how the regional flora and fauna interact with each other and with the physical environment. Gradually I became aware of the patterns of life in the Rockies. I learned the interconnectedness of this wondrous land.

When I think of the Rocky Mountains, I cannot help but see them as the backbone of North America, dividing the continent into two. East of the Rockies, all streams flow into the Atlantic Ocean. West of the Rockies, the streams run to the Pacific. Rising up after the dinosaurs disappeared some 65 million years ago, the Rockies extend from northeastern British Columbia in northern Canada to central New Mexico in the southern United States, a distance of more than 2,000 miles (3,200 km). On the east, the Rockies are bordered by the Great Plains, a sea of grassland washing up against the forested slopes of the mountains. On the west, the Rockies meet the more complex topographies of the Great Basin and the Rocky Mountain Trench.

The Rockies are huge. So different are the mountains of Canada from those of New Mexico, for example, that we commonly divide the Rocky Mountains into three sectors: northern, central, and southern. The northern sector of the Rockies lies in

NIWOT RIDGE LONG TERM ECOLOGICAL RESEARCH PROJECT, UNIVERSITY OF COLORADO

Aerial view of the Continental Divide in the Colorado Front Range

northern Idaho, western Montana, southwestern Alberta, and southeastern British Columbia. The central sector of the Rockies includes ranges in southern Montana, eastern Idaho, northeastern Utah, and western Wyoming. The southern Rockies extend from southern Wyoming to central New Mexico, a region that includes the Rockies's broadest and highest mountain terrain.

The Rocky Mountains lend themselves to superlative statistics. But facts and figures do not do justice to these mighty mountains. They are far more than just tall piles of rock. They are home to an amazing variety of living things. The Rockies form an archipelago of cool, moist habitats for plants and animals. Along the way, most of the upland regions of the Rockies are surrounded by grasslands. These range from cool prairies in Alberta to hot desert grasslands in New Mexico. Quite different from the neighboring grasslands, the Rocky Mountains represent a long string of habitat islands, occupied by plants and animals adapted for cooler climates.

At the tops of the high peaks in the Rockies, there are islands of tundra sitting atop a coniferous forest zone that climbs up the side of the mountain. These habitat islands are far smaller and more isolated than the sometimes broad bands of spruce, fir, and pine forests at lower elevations in the Rockies. They have close affinities with the arctic tundra that clothes the northern edge of North America, and many species of arctic plants and animals have small, isolated populations on mountaintops in the Rockies. The climatic conditions in the two widely separated regions are similar in many ways. Both the arctic and alpine tundra regions have short growing seasons separated by cold, long winters.

Alpine tundra near Trail Ridge Road, Rocky Mountain National Park, Colorado

Learning the natural history of an entire region takes us into the realms of botany, zoology, ecology, geology, geography, and climatology. I have lived nearly my whole life in the Rocky Mountain region, and I have been observing nature and doing ecological research there for more than twenty years. Yet I must draw largely on the work of others to tell the full story. The mountains still hold many secrets, yet so many fascinating things have been discovered it is an easy task to fill these pages.

To try to understand the current biological communities armed only with data obtained from modern studies would be as difficult as trying to understand a long novel by reading only the last page. My discussions of the natural history of the Rocky Mountains will include many references to things that happened thousands of years ago, the events that set the stage for what we see today. If people lived as long as bristlecone pine trees (i.e., several thousand years), our ecological theories would be very different.

My aim in this book is to bring all of my knowledge, and that of unnamed others, to you. I discuss the geologic history, physical environment, flora, and fauna of the Rockies, and I hope to impart a sense of wonder as I go. But I would be remiss to quit there, for the natural history of any region is not just the sum of what has happened up until now or a snapshot of the modern ecosystems frozen in time. Rather, natural history is an ongoing process, with a future as well as a past. Many ecosystems of the Rockies are in peril, and we must work for strong conservation measures to preserve them. It seems to me that some natural parts of the Rockies should remain unspoiled and that we should try to rehabilitate the regions that have been degraded by human effects. Quite simply, the future of the Rockies rests in our hands.

Selected References

Benedict, A. D. 1991. *A Sierra Club Naturalist's Guide to the Southern Rockies.* San Francisco: Sierra Club Books.

Gadd, B. 1995. *Handbook of the Canadian Rockies.* 2d ed. Jasper, Alberta, Canada: Corax Press.

Halfpenny, J. C., and R. D. Ozanne. 1987. *Winter: An Ecological Handbook.* Boulder, Colo.: Johnson Books.

Knight, D. H., 1994. *Mountains and Plains: The Ecology of Wyoming Landscapes.* New Haven, Conn.: Yale University Press.

Mutel, C. F., and J. C. Emerick. 1992. *From Grassland to Glacier: The Natural History of Colorado and the Surrounding Region.* 2d ed. Boulder, Colo.: Johnson Books.

Pole, G. 1992. *Canadian Rockies.* Vancouver, Canada: Altitude Publishing.

2

Mountain Building

Those of us who have grown up in western North America take rocks for granted. We are used to seeing big rock walls in canyons and road cuts. We walk over massive slabs of sandstone as we hike in the hills, and we visit national parks set aside for their beautiful rock formations. Although we have become accustomed to these geological wonders, visitors from other regions are astounded by them. In many parts of the country and the world, the geology is all covered up with vegetation. In the Rockies, you can actually *see* the geology. Why are the Rockies different? First, the Rockies are younger than many other mountain chains: They began to form only seventy million years ago. By way of comparison, the Appalachian Mountains began forming about 450 million years ago. If the entire history of this planet were condensed into a single day of twenty-four hours, the Rockies began rising after 11:30 PM. The Appalachian Mountains, the old timers of North America, were already old and wearing down by the time the Rockies were beginning to rise.

Although young from a geological perspective, earlier events contributed to the shaping of the Rocky Mountains. About 300 million years ago, a mountain chain called the Ancestral Rockies rose out of shallow seas. When these mountains eroded (about 260 million years ago), they left behind huge piles of sediment that formed sandstones that can still be seen today (for instance, along the eastern flanks of the Rockies in Colorado). Among these piles are the red sandstones that characterize the Fountain Formation sandstones that were tilted from their original horizontal position to nearly vertical positions along the Front Range in Colorado. These huge rock slabs were "stood up" as younger rocks rose to form the Rocky Mountains as we know them today.

The next major chapter in the story of the Rocky Mountains was the development of a huge, elongated basin, caused by the weight of great thicknesses of sedimentary rock that built up over millions of years. This basin extended from the Gulf of Mexico region to Alaska. It began forming during the Jurassic period (about 160 million years ago) and became a continuous seaway early in the Cretaceous period (about 130 million years ago). Streams carried sediments ranging from clays to gravel from upland regions surrounding the basin. Limestones were also laid down at certain times. The weight of the accumulating sediments caused the basin to keep sinking, providing space for additional sediment. This process continued for about 100 million years. Sedimentary rock layers as much as 30,000 feet (9,146 m) thick were laid down in some regions. These rock layers were the raw materials that would soon be used to build mountains.

Mountain building began with the folding and faulting of the rocks along the elongated basin. This huge deformation process began about 70 million years ago. As geological forces compressed the sedimentary rocks, the deformation of rock layers became more widespread and intense. Narrow upfolds of rock rose through the horizontal sedimentary rock beds that had been deposited much earlier. As groups of blocks rose, other blocks dropped to lower elevations. These downfolds sank between the uplifted blocks and became filled with sediments eroded from the rising rocks. The deformed layers that were forced upward formed mountains. This mountain-building episode lasted more than thirty million years, finally coming to an end about thirty-seven million year ago. Huge stresses developed in the crust as the layers of rock were contorted, squeezed, and stretched. Those stresses created many rock fractures, and molten rock, or magma, oozed up through these cracks and melted the rocks near the top of the crust. In some regions where cracks in the crust opened up, mineral belts formed. Ore minerals such as gold, silver, copper, lead, zinc, and iron sulfides oozed up from greater depths in the Earth's crust. The ore minerals crystallized and filled the cracks as veins. Veins rich in gold, silver, and other valuable

NANETTE ELIAS

View of the Flatirons, Boulder, Colorado. These slabs of ancient sedimentary rock were thrust up and tilted as the Rocky Mountains grew, beginning sixty-five million years ago.

metals are scattered throughout the Rockies, but the richest and most concentrated mineral belt occurs in the Rockies of Colorado.

Throughout western North America, the average elevation of the landscape is 4,000 to 7,200 feet (1,500 to 2,200 m). The mantle of the Earth (the region beneath the crust) was previously thought to be essentially uniform in thickness throughout the world, but recent research suggests that the mantle, like the crust above it, varies in thickness from region to region. The peaks of western North America may owe their height to increased warmth and "buoyancy" of the mantle beneath them. Warmer layers of the mantle are more buoyant, allowing the Rockies to "float" above the Great Plains.

Geology of the Northern Rockies

The mountains of western Alberta, eastern British Columbia, western Montana, and Idaho make up the northern Rockies. These mountain ranges represent many different rock types and land forms.

The Rocky Mountain Trench separates the eastern front of the northern Rockies in Montana from the western portion. This trench is actually a large block that has sunk between two faults. West of the trench granites dominate the high country. The granites formed as magma cooled on the surface of the rising mountains. In contrast to this, the eastern front of the Rockies in Montana are capped by sedimentary rocks that were pushed up along faults.

The geological history of the Rocky Mountain region of Montana began with the deposition of sediments roughly one and a half billion years ago. At that time, sandy and muddy sediments began accumulating in the western third of Montana, as well as in adjacent regions of Idaho and British Columbia. This sediment accumulation continued for about 600 million years. These deposits persist today as the Belt formations of western Montana, so-called because they were first studied in the Belt Mountains. Geologists know that these rocks are Precambrian in age because they contain only fossils of primitive plants, lacking any traces of animal life. In northwestern Montana, the Belt formation rocks contain many dikes and sills made of a black igneous rock.

The Belt formation deposits of Montana are covered by layers of sedimentary rock laid down in shallow seas, about 570 to 240 million years ago. Some 240 to 65 million years ago, most of Montana remained near sea level. Beginning about 100 million years ago, the crust under Idaho and Montana began to bulge as magma rose upward. The Earth's surface continued to bulge for millions of years, and the layers of sedimentary rock at and near the surface began to break up and slide off this bulging region. In some regions, the upper continental crust broke off all at once and moved eastward.

Throughout northwestern Montana and southern Alberta, layers of sedimentary rocks appear to have peeled off the bulged crust in a series of slabs as much as several thousand feet thick. One of the prime examples of this phenomenon can be seen today in the Lewis Thrust in northern Montana and southern Alberta. Blocks of sedimentary rocks were thrust up along a 280-mile-long (452-km-long) front, from southern Alberta through Glacier National Park, Montana. Precambrian rocks were thrust as much as 40 miles (64 km) eastward, up and over the tops of younger rocks.

CORAL CORPORATION

Mt. Rundle, Banff National Park, Alberta. Note the layers of sedimentary rock at the top of the mountain. Softer rock layers have eroded back, leaving the harder layers standing out from the surface.

For the past thirty years or more, geologists have been debating just how this happened. Initially, overthrusting was held as the most plausible mechanism to explain this massive repositioning of rock beds. In this scenario, the Precambrian rock beds were thrust up over the Cretaceous beds. During the past twenty years, however, a new theory has been put forward, which invokes gravitational sliding as the mountain-building mechanism. In the gravitational-sliding theory, a huge slab of crust (containing the Cretaceous rocks) gradually slid down the plane of a fault, eventually be-

Highest Peaks in the Rocky Mountain Provinces and States (Listed North to South)

Name	State or Province	Elevation above Sea Level
Mt. Robson	British Columbia	12,050 ft (3,954 m)
Mt. Clemenceau	British Columbia	11,146 ft (3,657 m)
Mt. Goodsir, South	British Columbia	10,857 ft (3,562 m)
North Twin	Alberta	11,332 ft (3,718 m)
Mt. Alberta	Alberta	11,031 ft (3,619 m)
Mt. Forbes	Alberta	11,000 ft (3,609 m)
Granite Peak	Montana	12,799 ft (4,199 m)
Mt. Woods	Montana	12,661 ft (4,154 m)
Koch Mountain	Montana	11,286 ft (3,703 m)
Borah Peak	Idaho	12,662 ft (4,154 m)
Leatherman Peak	Idaho	12,230 ft (4,012 m)
Hyndman Peak	Idaho	12,078 ft (3,963 m)
Gannett Peak	Wyoming	13,804 ft (4,529 m)
Grand Teton	Wyoming	13,770 ft (4,518 m)
Fremont Peak	Wyoming	13,730 ft (4,505 m)
Mt. Elbert	Colorado	14,433 ft (4,735 m)
Mt. Massive	Colorado	14,421 ft (4,731 m)
Mt. Harvard	Colorado	14,420 ft (4,731 m)
Wheeler Peak	New Mexico	13,161 ft (4,318 m)
Baldy Peak	New Mexico	12,623 ft (4,141 m)
Costilla Peak	New Mexico	12,580 ft (4,127 m)

coming lower in elevation than the older Precambrian rocks. Later the whole mass of rocks were uplifted together and tilted toward the west.

The Lewis Thrust Fault, whatever its origins, created some spectacular mountains in both Glacier National Park, Montana, and Waterton Lakes National Park, Alberta. The relatively soft beds of Precambrian rock that form the high country in this region alternate with more durable layers, creating a rugged beauty unlike that seen in granitic mountains.

The mountain-building process that began about ninety million years ago in Montana slowed to a near standstill about seventy million years ago. Then about fifty million years ago, volcanic mountains formed a series of chains ranging from the Cascades in the Pacific Northwest to the Gallatin and Absaroka Ranges of northern Wyoming and southern Montana. In central Montana, the Boulder batholith came to life again. Volcanic eruptions in south-central Idaho built volcanic peaks in that region at the same time. Deposits of sedimentary rocks, volcanic ash, and coal continued to form during the remainder of the Tertiary, but the great period of mountain building had ended. The last movement in this geological symphony was the sculpting of the Montana Rockies by glacial ice. This glacial reshaping of the landscape is perhaps best viewed today in Glacier National Park.

At about 1,545 square miles (4,000 km^2), Glacier National Park is not only large but also full of rugged terrain and scenic splendor. As might be expected, nearly all

St. Mary's Lake, Glacier National Park. The lake valley was repeatedly scoured by glacial ice, making it very deep and U-shaped.

of the physical features in the park have been chiseled, gouged, engraved, or trimmed by glacial ice. For instance, Lake McDonald owes its shape and size to the repeated movements of Pleistocene glaciers between the massive shoulders of Howe Ridge on the northwest and Snyder Ridge on the southeast. Sonar studies of the sediments in the lake show that the bedrock beneath it has been carved into a steep, V-shaped notch, about 885 feet (270 m) below the level of the lake. The series of lakes in the northwestern sector of the park are all oriented in this northeast–southwest direction. Logging, Quartz, Bowman, and Kintla Lakes are the products of glaciers that once flowed slowly down their respective drainages. The mountain valleys that wind their way between the high peaks are U-shaped, the product of glacial scouring.

About 18,000 years ago, the western edge of the Laurentide Ice Sheet advanced close to the east side of the Rocky Mountains in Montana, with a lobe of ice protruding almost as far south as the location of Great Falls. In the west, the buildup of glacial ice in the mountains of British Columbia formed the Cordilleran Ice Sheet, which butted up against the Laurentide Ice Sheet in western Canada and flowed south in Montana to the site of Flathead Lake. The two ice sheets covered more than 6 million square miles (16 million km^2) of North America and contained a third of all the ice bound up in the world's glaciers during the last glaciation.

The region now known as Glacier National Park was uniquely situated to record the events of the last glaciation because it was hemmed in by the two great ice sheets. In addition, mountain glaciers in the park region built up enough ice to spill

Deep U-shaped valleys in Glacier National Park were scoured out by glacial ice. View from Logan Pass, looking east.

down several drainages on to the adjacent lowlands. Glacier National Park is dominated by two mountain ranges. The line of these ranges trends northwest–southeast. Many small glaciers still cling to cirque walls near the tops of these mountain ranges.

About 20,000 years ago, only the highest ridges and peaks in the park were free of ice. The remainder of the park was a sea of streaming glaciers and ice fields. Glaciers that flowed from the western side of the Lewis Range south of Lake McDonald merged with ice from the southeastern flank of the Flathead Range. This ice floe, combined with other mountain glaciers southeast of the park, formed a large body of ice around the southwestern corner of the park. This body pushed up and over the low divide at Marias Pass and resulted in the formation of Two Medicine Glacier. This glacier was also fed by ice flowing from the southern end of the east slope of the Lewis Range. Farther north along the eastern slope of that range, mountain glaciers flowed out onto the plains to form the Cut Bank and St. Mary Glaciers. Ice from the northwestern flanks of the Lewis Range and the northeastern flank of the Livingstone Range flowed north into Canada.

Glacier National Park is the only U.S. national park in the Rocky Mountain region with a climate that allowed for continuation of substantial glaciers into the present day. The climate is favorable for glacier persistence because the park is situated far enough north and its mountains are high enough that they stay cool in summer. Another factor favoring glaciers is that the mountains in the park capture significant precipitation from moist Pacific air moving inland. The west side of Glacier National Park catches the lion's share of Pacific moisture. The difference between the wetter climate of the western slope and the drier climate of the eastern slope affects the modern-day vegetation of the park, but it also made quite a difference in the size, shape, and location of glaciers in the past.

The impressive glaciers that existed in northwestern Montana at the turn of the twentieth century, along with the impressive mountain scenery and wildlife, led to the creation of Glacier National Park. Not many people realize that when the area was set aside as a national park the glaciers there were in the process of thinning and melting back from their extent in about 1850, when they were bigger than they had been in the past 10,000 years. Thus the glaciers were far more impressive in the nineteenth century than they are today, but the nineteenth-century glaciers were kind enough to leave us some proof of their size and shape, in the form of lateral and terminal moraines. Several of these moraines, mounds of glacial debris pushed to the sides and foot of the glacier, have been dated through the use of tree rings. The age of trees growing within the area from which the glaciers had retreated indicated their maximum position during the "Little Ice Age," a time of relatively cold climate, from about 1400 to 1900 AD. The ice was substantial for many decades thereafter. Glaciers flowed down mountainsides, spilled over ledges, and choked lakes with icebergs. Where the ice flowed over steep terrain, it formed deep, dangerous crevasses. The bedrock beneath the glaciers was scoured, polished, and grooved as ice and rock debris ground their way downslope.

The Canadian Rockies, like their sister ranges in Montana, display massive beds of tilted sedimentary rocks at their tops. Fossils, such as ancient corals, are found at high elevations in the northern Rockies. The entire Canadian Rockies apparently moved northeastward along a large fault. The southwestern parts of the Canadian

Rockies contain the oldest layers of rocks, whereas the eastern foothills of the Rockies in Alberta contain the youngest layers.

The formation of the Northern Rocky Mountain Trench in British Columbia took place when blocks of the Earth's crust moved horizontally along a fault line. This trench contrasts with the Rocky Mountain Trench farther south, which was produced by an up-and-down movement of blocks as opposed to a side-by-side movement of blocks, which was seen in the Northern Trench. West of the Northern Trench, British Columbia has been moving in a northwesterly direction. Geologists estimate that this part of the continent has shifted 250 to 450 miles (400 to 750 km) northwest along this fault, in relation to the rest of North America. This is a slow process that has so far lasted tens of millions of years.

About two million years ago, the Earth's climate began to cool substantially, and a new geological era began, a more recent one that is characterized by ice ages, or glaciations. This is the Quaternary period, during which there have been at least seventeen major glaciations. The Quaternary period is divided into the Pleistocene Epoch, the time of the glaciations that lasted from 2.5 million years ago to 10,000 years ago, and the Holocene Epoch, the past 10,000 years, including the present. During the Holocene there have been no major glaciation events, but most interglacial warm intervals such as this have lasted only 10,000 to 12,000 years. From a geological standpoint, our brief moment in the Sun may be over quite soon. During the Pleistocene glaciations, ice sheets formed in the arctic regions, then the ice margins spread southward, eventually reaching the mid-latitudes. In the Rocky Mountains, glaciers grew in the highlands, then tongues of ice flowed downhill to the valleys, sometimes filling them for many miles. Glaciated mountain landscapes are easy to spot, because glacial ice has scoured, gouged, carved, punched, and bulldozed its way through the mountains, creating characteristic features and land forms. Glacial ice streams inexorably forward. The ice margin is like the front of a conveyor belt, carrying pulverized rocks and debris to the outer edge while the ice that is in contact with the ground is busy gouging out more material. If the sediments being carried along by the ice are mostly sand and silt, they may polish the underlying bedrock like sandpaper smooths a rough board. Ice that carries a load of pebbles and cobbles will act like an engraving tool, scratching deep grooves in the bedrock that lies in its path. Mountain valleys that were V-shaped before the glaciers ended up U-shaped, as their side walls were eroded back by the scouring action of advancing ice. These deep, steep-walled valleys are the hallmark of mountain glaciations. Inch by inch, layer by layer, the glaciers carved the Rockies into the landscapes we see today.

The Quaternary glaciations scoured the landscapes of the Canadian Rockies, forming typical glacial land forms, such as U-shaped valleys, horns (pyramid-shaped peaks), and hanging valleys. Glacial ice built up in the highland regions throughout the Canadian Rockies and flowed downslope, getting progressively deeper, until a more-or-less continuous ice sheet developed. This huge body of ice, called the Cordilleran Ice Sheet, swallowed up most of the Canadian Rockies during the last glaciation, about 18,000 years ago. As in Glacier National Park, Montana, only the highest peaks remained above the ice sheet. By 10,000 years ago, most of this ice had melted, but the Canadian Rockies stand poised on the verge of another glaciation, as demonstrated by the glaciers that persist on many mountaintops and ridges, including the Columbia Icefield, the largest ice field remaining in the Rockies today. This ice field gives us some unique insights into how glaciers work in the northern Rockies, both past and present.

The Colombia Icefield is approximately 125 square miles (325 km^2) in area. The size of the ice field is controlled by four main factors: elevation, latitude, topography, and climate. The ice field lies at about 9,850 feet (3,000 m) elevation, just north of 52° N latitude. It occupies a large alpine region ringed by high mountains.

A great deal of moisture is needed to build and maintain such a large ice field. There is a major valley opening to the southwest, channeling Pacific moisture carried by westerly winds up a vertical rise of 7,500 feet (2,200 m) right onto the center of the ice field. The ice field receives an average of 32 feet (10 m) of precipitation per year; most of this falls as snow, and little of it melts from year to year. The maximum known thickness of the ice is 1,197 feet (365 m), although it probably averages about 328 feet (100 m) thick. The ice field feeds several large glaciers, including the Athabaska, Bryce, Castleguard, Dome, Kitchener, Saskatchewan, and Stutfield Glaciers. Moreover, there are dozens of unnamed glaciers in the vicinity of the Columbia Icefield, as elsewhere in the Canadian Rockies.

Ice flows slowly from the ice field, feeding the glaciers that flow down valleys. It takes about 150 years for ice to travel from the outer edge of the ice field down to the end of the Athabaska Glacier. Ice that formed at the center of the ice field may take as long as 800 years to reach the foot of a glacier. Ice movement in the northern Rockies glaciers averages about 50 feet (15 m) per year on the flatter regions. In steep terrain, the movement of ice is more than 400 feet (125 m) per year. Although the glaciers in the Canadian Rockies have retreated somewhat since their late Holocene maximum during the nineteenth century, they have managed to maintain most of their ice mass. This is in sharp contrast to the glaciers in Glacier National Park, Montana, where most of the nineteenth-century glaciers have either melted completely or are down to less than 5 percent of their former ice volume.

Geology of the Central Rockies

The central Rockies are geologically diverse and have mixed origins. Precambrian rocks are extensively exposed in the Bighorn Mountains, the Beartooth Mountains, the Tetons, and the Wind River Range of Wyoming. The ranges near the border between Wyoming and Idaho are the result of rock formations that have thrust over adjacent blocks of rock.

One of the most geologically active regions in western North America is the Yellowstone region of northwestern Wyoming, a region that deserves some special attention. Standing next to a boiling hot spring or erupting geyser, it takes little imagination to realize that Yellowstone is a geological "hot spot." Most geological features change so slowly that they appear unchanging to human eyes. In contrast, the geothermal features of Yellowstone are very dynamic. Every few years, new hot springs or fumaroles appear, sometimes triggered by local earthquakes. Other hot springs dry up, and the frequency and size of geyser eruptions change. There are few places on the planet where the Earth is so energetic. To understand what happens in the geyser basins at Yellowstone, we need to delve deep beneath the Earth's surface. The Earth's crust serves as a layer of insulation, protecting the surface from the enormous heat of the planet's interior or mantle. The mantle layer is thousands of miles thick. It is mostly solid, but in some small regions it is hot molten rock, or magma. For reasons that geologists are still struggling to understand, magma builds up in pockets beneath the crust in some regions. Where these buildups occur, the magma pools are like blisters beneath the skin, which push up and thin the stretched crust. The heat of the magma melts the lower layers of crust rock, thinning the solid crust layer even further.

Yellowstone sits over one of these magma chambers. As a result, the Earth's crust in the Yellowstone region is only about 3 miles (5 km) thick, one tenth of the Earth's average crust thickness. As might be expected, a crust under this kind of stress is riddled with cracks as it is stretched and thinned over a growing cauldron of magma near the surface. Water from rain and meltwater from snow seep down into these cracks, where they become superheated. As the heated water builds up in the cracks, it expands and is forced up and out at the surface. Depending on the shape and size of the "plumbing" beneath the Earth, that water escapes as steam (in fumaroles), as boiling pools of water (in hot springs), or in a mixture of steam and water under tremendous pressure (as geysers). Geysers erupt in fits and spurts because each eruption releases part of the pressure in their "plumbing," and it takes time for that pressure to build again.

Some hot springs, such as Yellowstone's Mammoth Hot Springs, percolate up through bedrock that contains minerals that are readily dissolved, notably the calcium carbonate in limestone. The calcium carbonate dissolves in the hot water beneath the surface, then precipitates out of the water as it cools when it reaches the surface, forming terraces.

The Yellowstone region was covered by an ice cap during the late Pleistocene Epoch, yet even when the landscape was ice-covered, volcanic activity continued. As you can imagine, when red-hot lava meets glacial ice some interesting things happen. One such event was the formation of Obsidian Cliffs, south of Mammoth Hot Springs. At this locality, rhyolite lava flowed about 155,000 years ago. When it came in contact with glacial ice, it cooled so rapidly that the normal crystallization process did not take place. Instead, the lava turned to obsidian, a rare form of glass that has no crystalline structure.

In various parts of Yellowstone, the interaction of receding ice and hot springs created some interesting landscape features. In the vicinity of Mammoth Hot Springs, the edge of the glacier melted erratically, littering the landscape with large, isolated blocks of ice. These giant ice cubes remained long enough to be buried by sediments left behind by the receding ice margin. When the ice blocks finally melted, they left a series of alternating depressions and small hills, called a "kettle and kame landscape." Just south of Mammoth the cone-shaped feature called Capitol Hill and others nearby probably formed when glacial sediments accumulated in holes and embayments melted in the ice by the hot springs. Kettle and kame topography is common throughout the Yellowstone region, wherever blocks of ice were left stranded by receding glaciers.

The effects of the last glaciation can be seen throughout Yellowstone National Park and adjacent areas. The Yellowstone terrain visible today was shaped by a unique combination of forces: fire and ice. The fiery hot magma just beneath the surface has shaped the landscape through caldera explosions, lava flows, and geothermal features. Glacial ice scoured some regions and draped others with debris. On a geological time scale, Yellowstone is a dynamic, ever-shifting landscape, yet to anyone who visits it feels timeless.

Geology of the Southern Rockies

One of the prominent features of the southern Rockies are broad open areas partially or completely surrounded by mountains. These areas are called "parks" by geologists. The Laramie Basin in Wyoming and North, Middle, and South Parks and the San Luis Valley in Colorado are examples of parks.

Parks are structural basins that formed as blocks in the Earth's crust moved down through folding or faulting. They dot the landscape in places where rocks softer than those in the surrounding mountains are found. The soft rocks erode and the parks are formed. Geologically, the southern Rockies are a mixed bag. In the Colorado Front Range and the Medicine Bow Mountains of Wyoming, more than 600-million-year-old rocks are widely exposed. Indeed, old rocks such as granite dominate many regions. This is because granitic magma pushed up through other rocks more than 600 million years ago. One of the largest of these piles of granite is Pikes Peak. Some of the mountain ranges in the southern Rockies are composed mainly of upturned sedimentary rocks. Among these are the Mosquito Range west of South Park, Colorado, and the Sangre de Cristo Range in southern Colorado and northern New Mexico. A third group of mountains in this region was formed by volcanic activity. The San Juan Mountains of southern Colorado and the Jemez-Nacimiento Mountains of New Mexico were formed mainly by volcanic activity between sixty-five million and three million years ago. One of the most noteworthy geological features associated with volcanic activity is the Valles Caldera on the western flank of the Jemez Mountains in northern New Mexico. More than a million years ago, a volcano exploded, blowing out an amazing 50 cubic miles of debris. The walls of the volcano collapsed into a huge basin, called a caldera, which is 15 miles (24 km) in diameter.

The highest range of mountains in the southern Rockies is the Sawatch Range. It is made up of Precambrian rocks that have been extensively intruded on by magma flows, between forty million and twenty-five million years ago. Mt. Elbert, Colorado, the tallest mountain in the Rockies, is an example of a mountain formed by this process. Mt. Elbert crowns the Sawatch Range at 14,431 feet (4,400 m). It is one of fifty-four peaks in the southern Rockies that rise more than 14,000 feet (4,267 m) above sea level.

Not all the mountains of the southern Rockies were formed by uplifted blocks. Volcanic activity gave birth to many mountains. In some regions, great cracks in the crust allowed molten rock to come to the surface, creating lava flows and volcanoes. Between 25 and 40 million years ago, volcanic activity was intense in the San Juan Mountains, the Elk Mountains, and the Sawatch Range, all in southern Colorado. Deposits of volcanic ash thousands of feet deep blanketed much of southwestern Colorado as these volcanoes erupted. Toward the end of this fiery phase of vulcanism, massive flows of black basalt poured out of fissures in the Earth and flowed down the Rio Grande drainage in the San Luis Valley of southern Colorado and northern New Mexico. Two of the most impressive peaks left from this volcanic period are the

Spanish Peaks of southern Colorado. The surrounding landscape has eroded since then, revealing vertical slabs of volcanic rock that project away from the peaks. The Spanish Peaks themselves have eroded considerably since their formation; they were once much taller than they are today.

The history of glaciation is not as well worked out for the southern Rockies as it is for the central and northern Rockies. Most geologists think that glaciation in the southern Rockies was not as spectacular as glaciation in the central and northern Rockies. The buildup of glaciers depends on two factors: (1) There must be sufficient snowfall; (2) the snow must not melt away during the summer months. In other words, a region must be both cold enough and wet enough to sustain a large snowpack that persists for decades. The increasing weight of the snow eventually causes ice to form from the compressed snow crystals. When a sufficient mass of ice accumulates, it spills out of the high mountains and flows down valley, giving birth to a glacier. In the southern Rockies, summer temperatures have been cold enough to keep the snowpack from melting away, but there just was not enough snow to build really big glaciers.

We know little about the dozen or more glaciations that took place in the Rockies before the late Pleistocene, because the more recent glaciers obliterated most of the evidence left behind by the earlier glaciers. During the last glaciation, the largest glacier formed in Rocky Mountain National Park in Colorado was the Colorado River Glacier, which flowed from the alpine headwaters near La Poudre Pass in northern Colorado, merged with ice from the Never Summer Range in the national park, and

Spanish Peaks, southern Colorado. These massive peaks are volcanic in origin.

formed one large river of ice that crept down the Colorado River drainage. The ice may have been as much as 1,500 feet (450 m) thick near the heads of the glaciers in the park.

Most of the lakes in Rocky Mountain National Park are the product of glaciation. Several of the park's lakes, such as Bear Lake, occupy "kettle holes" that were formed as stagnant ice melted during the last deglaciation. The shores of these lakes are littered with boulders and other glacial debris. Other lakes in the park were formed as water-filled basins were scoured out of high mountain walls by glacial ice. These lakes include Blue Lake, Black Lake, Chasm Lake, Fern Lake, Odessa Lake, Shelf Lake, and Tourmaline Lake.

Peaks and Valleys

The forces that acted to push up the Rockies were many and varied throughout the long chain of mountain ranges. Blocks of rock moved up, down, and sideways. Huge bodies of molten rock forced their way up to the surface and reached towering heights in the southern Rockies, causing old seabed formations to weather off the top or slough off to the sides. Volcanic cones built up from layer on layer of molten rock. Vulcanism also spread deep carpets of ash and occasional mudflows in the Rockies. These formed layers of soft rock that erodes easily. To the north, older sedimentary rock layers were heaved up and over the top of younger rock formations, and the combination of geological forces also caused shifting blocks of rock many miles to the east. The result is that the northern and southern Rockies have quite a different appearance. In the south, the sedimentary rocks are mostly perched along the sides and bottoms of peaks.

The sedimentary rocks perched on top of the northern Rockies have weathered to produce some characteristic shapes that typify the highlands of this region. Names applied to these shapes are overthrust mountains, dogtooth mountains, sawtooth mountains, and castellated mountains. Overthrust mountains feature a tilted southwest-facing slope and a steep northeast-facing cliff. This shape illustrates the direction and angle of how the blocks of sedimentary rock beds were thrust up in a northeastern direction over other rock beds. Examples of overthrust mountains include Mt. Rundle, Endless Chain Ridge, and Sunwapta Peak, all in the Rockies of Alberta.

Dogtooth mountains formed where sedimentary formations were thrust nearly vertically during mountain building. They stand now as relatively resistant spires in landscapes where the surrounding softer layers have eroded away. Examples of dogtooth mountains in Alberta include Mt. Birdwood, Cinquefoil Mountain, and Mt. Edith Cavell, as well as Spike Peak in British Columbia.

Sawtooth mountains are characterized by long ridges that mark the upturned edges of thrust sheets. These ridges are set at perpendicular angles to the prevailing wind direction. Weathering has eroded gullies on their southwestern slopes to produce the sawtooth shape. Examples of sawtooth mountains include the mountains in the Colin, Queen Elizabeth, and Sawback Ranges of Alberta.

Castellated mountains typify the eastern main ranges in the Canadian Rockies. In these mountains, weathering-resistant layers of dolomite, limestone, and quartzite are separated by weak layers of shale. The more resistant rock layers form cliffs and the weak layers have eroded into ledges. These alternating layers give a layer-cake appearance. Unlike dogtooth and sawtooth mountains, the beds in castellated mountains remained nearly horizontal in the mountain-building process. Examples of castellated mountains include Mt. Amery, Castleguard Mountain, Castle Mountain, Pilot Mountain, Mt. Saskatchewan, and Mt. Temple, all in Banff National Park, Alberta.

Glacial Landscapes

Mountain building laid the foundation for the Rockies, but glacial ice had the last word in sculpting their shape into what we see today. Geologists estimate that at least seventeen glaciations, or ice ages, have occurred during the past 2.5 million years. It is likely that each of these glaciations buried parts of the Rockies in ice, although the actions of the most recent ice age tend to wipe out the evidence for previous glaciations, unless the older glaciers extended farther down valley than the more recent glaciers.

In all but the southernmost Rockies, glacial landscapes dominate the highlands. The forms are many, and so typical of mountain scenery that we do not give them a second thought. But if the Rockies had never been glaciated, they would look different. The peaks would be more rounded, the valleys would be V-shaped. The mountain slopes would curve gradually down to the valleys in gentle arcs. There would be few lakes, few precipices, and almost no waterfalls. You can imagine these unglaciated Rockies like a block of roughly hewn marble: pleasant to gaze at for a little while but certainly not fascinating. Glacial ice acted as a sculptor, shaping that marble into a masterpiece, full of dramatic angles, sweeping curves, and exquisitely carved details.

The sheer cliffs and sharp peaks of most of the Rockies owe their shape to sculpting by glacial ice. As ice flowed over bedrock, it eroded surfaces to different degrees, creating benches and stair-step valleys. Long, sharp mountain ridges are typical of the highest landscapes along the Continental Divide. They were formed when ice scraped along both sides of a ridge, leaving oversteepened slopes that come to a sharp crest at the top.

Some of the high, steep-walled valleys perched near the tops of many mountains in the Rockies remain filled with glacial ice. Where the glaciers have melted, the depression at the base is often filled with water, forming a lake. In the Canadian Rockies, glaciers persist on shaded north- and east-facing slopes. Sometimes the ice clings to the steep slopes at seemingly impossible angles, and I find myself wondering, "Why doesn't the ice just fall off the mountain?"

Where a larger valley was eroded by a large glacier and a smaller, higher valley was eroded by a tributary glacier, "hanging valleys" form. These glacial valleys have a mouth that is perched at a relatively high level on the steep side of a larger glacial valley. At the mouths of these hanging valleys you will often find spectacular waterfalls, such as Bridal Veil Falls in Banff National Park, Alberta.

Pyramid-shaped peaks called "horns" were shaped by glaciers grinding their way along three or more sides of a mountain. Examples of horn mountains include Wetterhorn, Matterhorn, Longs Peak, and the Maroon Bells, all in Colorado; the Grand Teton in Wyoming; Flinsch Peak, Fusillade Mountain, Reynolds Mountain, and

Bridal Veil Falls in Alberta

Thunderbird Peak in Montana; and Mt. Assiniboine, Mt. Athabaska, Mt. Chephren, Mt. Fryatt, and Mt. Carnarvon in Alberta.

Lakes and Streams

Surrounded by parched lands, the Rockies are the source of life-giving streams that water much of western North America. The natural lakes in the Rockies were mostly created by glacial action. The glaciers gouged out basins large and small, then left natural dams in the form of terminal moraines at the downstream ends of valleys. One exception to this is Yellowstone Lake, in Yellowstone National Park. It sits in a depression formed by volcanic activity, but even that depression was scoured by glacial ice, changing its shape. Other large lakes owe their existence to ice-scoured basins dammed by rocks and debris from glacial moraines. These include Flathead Lake, Montana, and Jackson Lake, Wyoming.

Where glaciers are still active in the Canadian Rockies, local lakes and streams take on a beautiful turquoise-green color. This color is not just a reflection of the sky. Even under gray, cloudy skies the lakes are still turquoise green. Nor does the color come from any minerals at the bottoms of the lakes. Rather, it comes from the finely ground sediments produced by the glaciers, called rock flour, which is mostly silt. When silt is placed in water, it stays in suspension for hours or days. The suspended silt particles absorb wavelengths of light in the red to yellow range, and reflect back wavelengths of light at the blue and green end of the spectrum. So the

Lower Waterfowl Lake, Icefields Parkway, Alberta. The turquoise green color is derived from rock flour, finely ground sediments produced by nearby glaciers.

whole spectrum of light enters the water but only the blue and green colors of light are reflected back to the eye. It therefore does not matter if the sky is blue or gray, the lakes remain blue-green because of the reflectance properties of the glacial flour they contain.

Many of the Rockies's most beautiful waterfalls originate in areas carved by glacial ice. Some, like Bridal Veil Falls in Alberta, are slender ribbons of water, cascading down canyon walls below hanging valleys. Others, like the lower falls of the Yellowstone River near the Canyon visitors' center in Yellowstone National Park are raging torrents of water transporting thousands of gallons per second over rock ledges that are resistant to erosion.

In the northern Rockies where many glaciers are still active, mountain streams are choked with a heavy load of glacial debris. Such streams are constantly shifting across numerous channels as they make their way through valleys filled with glacial sediments. When the supply of sediment tapers off, streams settle down into a single channel. Over time, the course of channels in those streams not given to frequent flooding may meander back and forth, giving the stream many hairpin bends.

Floods send torrents of water downstream. These madly dashing waters follow the path of least resistance, which is usually straight downhill. The scouring action of floods cuts through meanders, straightening their course. Floods can be frequent events in the Rockies. Given the right circumstances, such as a heavy thunderstorm dropping more than an inch of rain on already rain-soaked ground, the mountain slopes can act as funnels, gathering large quantities of water at their tops and delivering walls of water down the narrowing drainages confined in mountain canyons. In recent decades, mountain floods in Colorado have killed hundreds of people and destroyed millions of dollars in real estate. When a wall of water 40 to 50 feet (12 to 15 m) high comes down a canyon, it can move boulders the size of houses. It is foolhardy to try to beat the water in a race down the canyon. The only way to be safe in such circumstances is to climb up the sides of the canyon, out of the water's path. Indeed, warning signs urging this procedure have been positioned along Colorado's canyon roads in recent years.

Climatology

Climate is a term used to describe regional atmospheric conditions over decades and centuries of time. The features of a region's climate include wind speed and direction, precipitation, temperature, and relative humidity. Climatic conditions are usually expressed in terms of monthly, seasonal, and yearly averages, using data compiled for thirty or more years. The term *weather*, on the other hand, is used to express atmospheric conditions at a given time and place. The climate of a region reflects the average patterns of weather conditions over many years. This differentiation is especially crucial in mountain regions, where average conditions (climate) cannot be used to predict local weather. The weather patterns of mountain regions are highly complex and change rapidly. For example, a warm, sunny summer morning can turn to a bitterly cold afternoon with 50-mile-per-hour winds and rain, sleet, or snow coming down in sheets.

In the Rockies, as elsewhere, the climate is an extremely important factor in the shaping of regional landscapes. The climate exerts strong, sometimes dominating controls on vegetation patterns, land forms, weathering of rocks, soil formation, and distribution of animal life.

There are three major climatic gradients in the Rocky Mountains. The first is elevation. The higher you go in the Rockies (or any place else), the cooler the air. The second climatic gradient is latitude. The farther north you go, the cooler the air. The third gradient affecting the Rockies is the precipitation pattern. Generally speaking, the western slopes of the Rockies receive more moisture than the eastern slopes, because the prevailing wind direction is from west to east.

Let's begin our examination of Rocky Mountain climate with the most obvious feature: elevation. High elevations in the Rockies are surrounded by air that is considerably "thinner" than at sea level. Anyone who has exercised in the Rockies can attest to this. The air contains less oxygen, because atmospheric pressure decreases with elevation. The thin air absorbs less energy from the Sun, and the heat that has been absorbed is quickly lost to the upper atmosphere after dark. As a general rule, for the southern and central Rockies, temperature decreases by about 1°F per 328 feet (about 0.6°C per 100 m). All other things being equal (which in reality they hardly ever are), Denver, Colorado (5,280 ft elevation, or 1,600 m) is considerably warmer than Niwot Ridge (12,277 ft elevation, or 3,743 m). The average July temperature at the weather station on Niwot Ridge is 46.6°F (8.1°C). The difference in elevation between the Niwot Ridge station and Denver is 6,997 feet (2,133 m). Average July temperature at Niwot Ridge should be 23°F (12.8°C) cooler than Denver, but it is actually 26°F (14.6°C) cooler than Denver, so the difference between these two sites is closer to 0.2°F per 100 feet (0.7°C per 100 m).

An exception to the "colder as you go higher" rule can be found in mountain valleys. These valleys are often colder than the slopes of neighboring mountains, especially in winter. This valley cooling is brought about by cold air drainage. Cold air is heavier than warm air, and it tends to sink down mountainsides and settle in valleys. Mountain valleys can be chilled quite severely in winter, especially if they are surrounded by high peaks. So places like Steamboat Springs, Colorado, have average January temperatures that fall well below what they should be, given their elevation. In fact, winter temperatures in these valleys are close to temperatures thousands of feet higher.

Latitude certainly reflects temperature but is not so easily worked into a simple formula. Denver, Colorado, is situated in the eastern foothills of the southern Rockies, at 5,280 feet (1,600 m) elevation and about 40° N latitude. The average July temperature at Denver during the past thirty years has been 73°F (22.7°C). In contrast, Calgary, Alberta, similarly situated in the eastern foothills of the northern Rockies but at 51° N latitude, experienced average July temperatures of 62°F (16.4°C). In some ways, this is not a strictly fair comparison, because Calgary is almost 2,000 feet (600 m) lower in elevation than Denver. Other sites in Alberta that occur at similar elevations to Denver are in the mountains, and mean July temperatures at these mountain sites are closer to 57°F (14°C).

One of the chief causes of temperature differences is the angle at which the Sun strikes the Earth. At lower latitudes (i.e., closer to the Equator), the Sun's energy strikes the Earth at almost a right angle, producing more atmospheric heating. At high latitudes, the Sun strikes the Earth at lower angles, producing less atmospheric heating. One of the more obvious ways in which latitude differences can be seen relates to treeline. *Treeline* is the upper limit at which trees grow. At this upper limit, the spruces and firs are not tall, upright trees but rather low, creeping shrubs. In the southern Rockies of New Mexico and southern Colorado, the treeline occurs at about 12,000 feet (3,650 m). In the northern Rockies of Alberta and British Columbia, the treeline is found at 4,920 feet (1,500 m). Ecologists believe that temperature is the principal control on upper treeline. The temperature regime controls such important factors as the length of the growing season, the time of snowpack melting in the spring, and the warmth of the air during the preciously short summer, when trees either succeed in growing and reproducing or fail.

Moisture Sources

Precipitation is the third critical climate factor in the Rockies. Moisture-laden air that comes to the mountains often arises over oceans and travels thousands of miles to the Rockies. The Rockies receive moisture from three sources of oceanic air. The first point of origin for moisture-laden air is the Pacific waters of Mexico. The San Juan Mountains of the southern Rockies are the greatest beneficiaries of this Pacific moisture source, although during the summer monsoon season this moisture may reach northern Colorado and even Wyoming. The second source of moist oceanic air is the Gulf of Mexico. The southern Rockies receive moisture from storm tracks moving northwest from the Gulf of Mexico. Sometimes the eastern slope of the southern Rockies gets the lion's share of this moisture, especially during March and April, when this region gets its heaviest snow storms. Most of the record-setting snow storms in this region have occurred in these circumstances. Indeed, the greatest recorded one-day snowfall in the United States occurred at Silver Lake, Colorado, in April 1921, when 6 feet, 4 inches (1.85 m) of snow fell in twenty-four hours.

The third source of moisture is cool Pacific air that travels south and east from the Gulf of Alaska and the coast of British Columbia. The central and northern Rockies receive most of the Pacific Northwest moisture. The northern Rockies are too far north to receive much moisture from either the Gulf of Mexico or the subtropical Pacific, and the southern Rockies rarely receive moisture from the Pacific Northwest.

The Rockies are sometimes affected by storm patterns driven by events in the tropics. Warming of sea-surface temperatures in the central Pacific creates currents in the atmosphere. This in turn generates large storm clouds over the tropical ocean, and the clusters of storm clouds generate heat in the mid-atmosphere that causes a repositioning of the jet stream over North America. These events take place roughly every six to eight years, and the subsequent changes in sea-surface temperatures prevent the deep, cold water near the western coast of South America from rising to the surface. This is the El Niño climate pattern. The normal winter storm patterns of western North America can be greatly disrupted by El Niño events, because the altered trajectory of the jet stream brings cold arctic air well south of its normal position.

Snow

Snow defines the winter landscape of the Rockies, changing everything almost overnight. At lower elevations, snow falls only during the coldest months, but on the mountaintops snow falls every month of the year, and snowbanks may persist for decades or longer. Mountain snows govern the life patterns of montane plants and animals for most of the year, keeping it subdued, hidden, or dormant. On the other hand, snowpack that accumulates in winter is the main source of life-giving moisture that makes the mountains livable for many species of plants and animals that are not adapted to the arid and semiarid lowlands that surround the Rockies.

Needless to say, not all snowflakes are alike. Not only are the crystals each shaped differently, they also come in a wide variety of forms. Snow crystals all start as water vapor in the atmosphere. As the vapor-laden air rises over the mountains it cools, and the air loses some of its ability to hold moisture. When air masses rise high enough in the atmosphere, they cool to temperatures below the freezing point of water (32°F/0°C). Minute particles in the atmosphere, such as dust, volcanic ash, or bacteria, form the nucleus around which water freezes to form snowflakes.

The newly formed snowflake is a hexagonal crystal, made up of as many as 100 million water molecules. The ultimate shape of the snowflakes that hit the ground depends on the atmospheric conditions under which they are formed and what happens to them on their descent to Earth. Large, elaborate crystals form in very moist air, such as is found in large, relatively warm clouds. Smaller, less elaborate crystal shapes form in colder, drier clouds that have less moisture to offer.

With each trip up and down through the clouds, the crystals grow. Eventually such masses of lumpy ice become too heavy to be easily lifted up to greater heights, and they fall. As individual snow crystals fall, they collide with other crystals; these bond together to form what we call snowflakes. Each snowflake has a unique history, beginning with the formation of many ice crystals in the atmosphere, followed by repeated bumping into other crystals on the way down, and the formation of groups of crystals into a cohesive flake that reaches the Earth. Each snowflake, then, is not an individual but rather a "committee of ice crystals," brought together haphazardly through a series of atmospheric collisions.

The things that happen to snow once it has fallen are even more complicated than what happens to the ice crystals as they fall from the atmosphere. Snow layers accumulate on the ground, with thick deposits following big storms and thin deposits following little ones. In the Rockies, the wind blows snow around, and windy days are often followed by bright sunny days. The Sun then melts the top layer of windblown snow, even if the air temperature remains below freezing. The melted snow layer sends liquid water down through the snowpack. If the snow is compressed deep enough and long enough, snow crystals break down. What was fluffy snow congeals into layers of ice, a process that can lead to glacier formation.

Even though we think of snow as just sitting up on the mountains, there is much going on beyond our gaze. The water frozen in snow crystals tends to change from solid snow to liquid to water vapor, and back again, ending as ice. As liquid water and water vapor move through the snowpack, they alter the snowpack's shape and density. Snow crystals generally do not cool much beyond their freezing point, and light, fluffy flakes trap many tiny pockets of air as they accumulate. The result is that a dense snowpack serves as an excellent layer of insulation, protecting the ground underneath from the extremes of winter cold. Warmth from the ground can actually creep up into the snowpack, making the bottom layers warmer than the surface layers. In spring, the Sun heats the upper layers and the temperature of the whole snow column, from ground level to surface, tends to be about the same.

These, then, are the basic elements of climate that dictate the composition and character of life in the Rockies. The plants and animals that inhabit the Rockies are subject to weather on a day-to-day basis. If weather patterns are such that it does not rain for six or seven weeks in a given summer, many plants will wither and die. Obviously, weather patterns are what make for "good" or "bad" seasons for biota and can leave lasting effects for days, weeks, or years.

Weather Patterns of the Rockies

Many years ago, a popular song waxed eloquent about "springtime in the Rockies." Actually, this phrase is somewhat of an oxymoron. The Rockies hardly ever have what would be recognized elsewhere as "spring" weather. The snow may get a little shallower and the temperatures rise during the daytime, but spring storms often dump large amounts of fresh snow in March, April, and May. In the northern Rockies, the only months in which snowfall is rare are July and August. The same holds true a thousand miles to the south. Only the lower elevations of the southernmost Rockies seem to be immune from late spring and early autumn snows. Skiers rejoice at this, whereas backpackers must either bide their time or strap on snowshoes to go hiking in the high country.

Summer rainfall in the Rockies often arrives in the form of afternoon thunder showers. These are mostly convection storms, spawned by warm air that rises thousands of feet in the atmosphere. As the Sun-warmed air pumps up to higher elevations, it cools off, and the moisture it contains condenses into those familiar, puffy, cumulus clouds. By mid-afternoon these clouds produce thunderstorms and scattered rain showers. By early evening, the solar energy required to drive the storms tapers off, and mountain skies generally clear. However, as the cumulus clouds drift eastward onto the plains, they can build in height, often reaching 30,000 feet (9,150 m) or more. There is tremendous energy trapped in these clouds, and the continued rising and falling of water vapor, raindrops, and ice at the various elevations within such clouds can lead to the buildup of hailstones that plummet to Earth, pummeling the plains east of the Rockies. In the worst-case scenario, these enormous black clouds spawn tornadoes that drop down to Earth and wreak havoc. Meanwhile, back in the Rockies, all is calm as night falls.

Violent thunderstorms that sweep out of the Rockies and onto the plains east of the mountains tend to be most powerful in the southern Rockies, because all of the energy required to make such storms comes from the Sun. Solar energy is more potent at lower latitudes, and thus has less effect in the northern Rockies. Therefore, thunderstorms are extremely rare in the Arctic but common in the mid-latitudes.

Cold-weather storms produce most of the moisture in the annual precipitation budget of many regions in the Rockies. In many regions of the Rockies, the high-elevation forests build the heaviest snowpack each winter. The same amount of snow may fall on the tundra, but this treeless, wind-swept region cannot hold on to the snow it receives, except in snowbanks that form in depressions. Much of the snow that falls on the tundra is blown down to the forests, where the trees reduce the wind speed to the point where snow falls from the air.

Soils

The soils of the Rocky Mountains are mostly thin, rocky, and poor. Soil scientists have developed a classification scheme for soils. This system is bewilderingly complex and full of tongue-twisting terminology, but what it shows us is that soils, like the plants and animals they support, are made up of unique combinations of parent materials (sands, silts, and clays derived from various types of rock) and living and decomposing life forms (decayed and decaying plant and animal bodies and microbes, mainly microscopic plants and animals, bacteria, and fungi). Soils also have a history. At the end of the last ice age, the soils of the Rockies had been stripped away in many places by glacial ice. The newly exposed ground received a load of essentially sterile sediment from the receding glaciers, and this load of sand, silt, and clay has slowly, over the past 10,000 years, become the mountain soils we see today.

In the Rockies, two main types of soils dominate the landscape. The first was called gray wooded soil in an earlier (easier to pronounce) classification, and has since become known as *Cryoboralf.* This type of soil exists in cool, humid environments. The other important soil type in the Rockies used to be called tundra or alpine soils, and is now called *Pergelic cryaguepts.* These soils exist in the high country, on the cold, wet, high divides and cirques above treeline. Let's examine these two main soil types in turn.

Cryoboralfs develop beneath conifer forests (spruce, fir, and pine) at elevations between 6,000 and 11,000 feet (1,830 and 3,350 m). In this region, annual precipitation averages 18 to 45 inches (460 to 1150 mm), and annual average temperatures range from 33 to 45°F (1 to 7°C). These *Cryobaralf* soils have distinct layers. At the top, they generally have an accumulation of partially decayed conifer needles and other plant debris. Conifer needles are slightly acidic, and these acids percolate down through the soils, creating slightly acidic soils well-suited to the growth of conifer forests (the trees create ideal conditions in the soil in which they grow). Just beneath the surface is a gray zone, followed by a brown zone that varies in thickness and may have quite a bit of clay in it. These soils cannot sustain enough vigorous grass growth to keep grazing animals well-fed because the soils lack sufficient organic matter and nutrients. However, grazing wildlife and livestock do consume new growth during spring. Likewise, these soils are too nutrient-poor to farm and occur in areas with a brief growing season and on slopes too steep for plowing. As poor as they are, however, these soils perform a vital role in retaining water and in controlling runoff from melting snow and summer thunderstorms. When the herbaceous ground cover and dead conifer needle layer over these soils is destroyed, such as during forest fires, these soils erode quite easily. One of the most common and devastating aftermaths of mountain fires occurs when thunderstorms drench the burnt region with rain. Massive mud slides often follow hard on the heels of these storms, as the charred soil erodes down the mountain sides.

Nutrient cycles are a key element that shapes the ecosystems of the Rocky Mountains. Mountain terrain constantly loses nutrients through soil erosion and stream flow. Once nutrients are lost in this way, they do not return, because they are carried downslope to other regions. However, nutrients can be recycled within an area for an indefinite period of time. For example, a nutrient may be used many times over by a succession of soil microbes, plants, and animals. Nitrogen, which is a vital element for plant growth, is commonly recycled within mountain ecosystems.

The annual nitrogen cycle in a typical Rocky Mountain forest begins with a three- to six-week period of melting snow and runoff, known as the "spring flush." During this time, nitrogen compounds are leached out of the soil, as they get mixed with water. Mountain regions also lose nitrogen through soil erosion, animal emigration, and the removal of nitrogen (denitrification) by soil microbes. Additions to the nitrogen nutrient pool of mountain regions also come in various forms. Nitrogen-rich dust falls out of the air, and nitrogen dissolved in rain and snow water falls in precipitation. Nitrogen also weathers out of rocks and is transported into a region by animals. Except in the high alpine regions, nitrogen flows into regions by surface and groundwater. Finally, soil microbes convert atmospheric nitrogen into forms useable by plants, including ammonium and nitrate. Studies in the lodgepole pine forests of Wyoming indicate that nitrogen accumulates there even in years of a heavy spring flush. Much of this nitrogen, about 90 percent, is held in organic matter in forest soils, whereas the living plants and animals hold only 4 percent of the forest's nitrogen. Plants obtain nitrogen from the soil organic matter and return it to the soil when they die. It appears that dead fallen trees are a major reservoir of nitrogen in the Rocky Mountains.

The soils of the alpine tundra, *Pergelic cryaguepts*, develop in regions of heavy snowfall and a very short growing season (the few weeks between summer snowmelt and autumn snowfall). Alpine soils support herbs and small shrubs, and these plants often form a thick mat at the top of the soil column. This upper layer of alpine soil is often almost black at the surface; it becomes more gray-colored as it deepens and ranges from moderately to strongly acidic. Alpine soils are extremely shallow; they often cover bedrock in the high country with only a few inches of soil. Because the high country was the last region to rid itself of glacial ice at the end of the last ice age, alpine soils are the youngest soils of the Rockies, and are often frozen for more than nine months of the year. Indeed, some alpine soils are permanently frozen (permafrost). The microbial activity that enriches most soils is greatly slowed down or excluded from these very cold soils, making alpine soils less developed than soils at lower elevations. Finally, these mountaintop soils are upstream from all other soils, allowing nutrients, minerals, and organic matter to move downslope with water.

The foothills and plains east of the Rockies have more productive soils than occur in the mountains. The lower elevation soils adjacent to the Rockies support relatively rich growth of grasses and forbs. Although they are not rich in organic matter (hence their relatively light brown color), they yield modest crops of grains and may support grazing livestock. However, these soils exist in a semiarid region where precipitation is critical to maintain these activities. As was observed in the 1930s, a few years of drought can turn these grasslands into dust bowls.

Water

The waters of the Rocky Mountains are an extremely important component of regional water supplies in western North America. The Colorado Rockies alone are the source of five major river systems: the Arkansas, Colorado, Rio Grande, San Juan, and the North and South Platte Rivers. The Yellowstone region of Wyoming contains the headwaters of the Snake and the Yellowstone Rivers. The Montana Rockies spawn the Missouri River. Farther north, the eastern slope of the Rockies in Alberta gives rise to the North and South Saskatchewan Rivers and the Athabaska River, and the western slope of the Rockies, in British Columbia, contains the headwaters of the mighty Columbia River (the third largest river by volume of water in North America), as well as the Fraser and Kootenay Rivers. The average annual flow of water in the Columbia River is about 110 billion gallons (465 billion l) of water per day. It is difficult to comprehend this quantity of water; one billion gallons is equal to a column of water 2,331 feet (710 m) high with a base the length and width of a football field. The Missouri River is somewhat smaller but still the fifth largest North American river, with an average daily discharge of about 50 billion gallons (211 billion l).

The Rockies pluck those billions of gallons of water from the moist air that flows over the mountaintops. In the high country, most of the precipitation falls as snow. Most of this snow does not melt before late May or June, and snowbanks in sheltered locations persist for many years. Eventually the water is released from the clutches of Old Man Mountain, and it makes its way down to the lowlands. Much of the lowland regions surrounding the Rockies are what the nineteenth-century explorers referred to as "the Great American Desert." These are regions that may receive only a few inches of precipitation per year, and without the mountain water that flows through them these lowlands would indeed be inhospitable to most life. The Colorado River is one of the most heavily used water sources in the arid Southwest today. The flow of the river near Lees Ferry, Arizona (just above the Grand Canyon), averages about 1.1 billion gallons (4.7 billion l) per day, but by the time the water reaches the river's mouth at the Gulf of California, under normal conditions the flow has been reduced to about 275 million gallons (1.2 billion l) per day, or just one quarter the volume of water flowing in the upper reaches of the river.

The Rockies water a huge region of North America. The resource, however, is not unlimited, and as more and more people live in the arid and semiarid West, the water resources of the Rockies are likely to be exhausted. As we begin the twenty-first century, water conservation will have to be a part of the daily lives of westerners.

Selected References

Alt, D., and D. W. Hyndman. 1986. *Roadside Geology of Montana.* Missoula, Mont.: Mountain Press.

———. 1989. *Roadside Geology of Idaho.* Missoula, Mont.: Mountain Press.

Benedict, A. D. 1991. *A Sierra Club Naturalists Guide: The Southern Rockies: The Rocky Mountain Regions of Southern Wyoming, Colorado, and Northern New Mexico.* San Francisco: Sierra Club Books.

Chronic, H. 1980. *Roadside Geology of Colorado.* Missoula, Mont.: Mountain Press.

———. 1987. *Roadside Geology of New Mexico.* Missoula, Mont.: Mountain Press.

Fritz, W. J. 1992. *Roadside Geology of the Yellowstone Country.* Missoula, Mont.: Mountain Press

Gadd, B. 1995. *Handbook of the Canadian Rockies.* 2d ed. Jasper, Alberta, Canada: Corax Press.

Lageson, D. R., and D. R. Spearing. 1988. *Roadside Geology of Wyoming.* Missoula, Mont.: Mountain Press.

Mutel, C. F., and J. C. Emerick. 1992. *From Grassland to Glacier: The Natural History of Colorado and the Surrounding Region.* Boulder, Colo.: Johnson Books.

3

Fossils

The Rockies house a treasure trove of fossils and have been one of the most important areas for paleontologists studying ancient life. Even though the Rockies are far-removed from modern oceans, they contain tremendous deposits of fossil sea creatures, because major parts of the Rocky Mountain region contain sediments that were originally laid down in shallow tropical seas. In terms of terrestrial fossils, some of the most important dinosaur fossil beds have been found in the Rockies, from Colorado to Alberta.

Earliest Fossils from the Rockies

For the first few billion years there was either no life on Earth or the traces of life were so faint that they have been lost to the fossil record. In the Rocky Mountains, the oldest known fossils date to about 1,700 million years ago, about three billion years after the formation of the Earth. Some of the oldest fossils in the Rockies are stromatolites that formed as mounds of calcium carbonate produced by marine algae in shallow tropical reefs. They are still found in shallow tropical seas today, so in their own way they have been very successful. Stromatolites are found in the Medicine Bow Mountains of southern Wyoming. Stromatolites can be seen at Stop 16 on the Going to the Sun Highway in Glacier National Park. In Canada, 1.3 billion-year-old stromatolites are found in the Siyeh Formation of Alberta.

The Cambrian period (570–500 MYA) saw an explosion of life forms, both in the seas and on land. Hard-shelled, jointed arthropods dominated the animal kingdom and left behind fossil impressions in the sands and muds of their day.

Today, arthropods are represented by insects, spiders, and crustaceans. Perhaps the most famous of the early arthropods was the trilobite group, the quintessential fossil of the period. But brachiopods and mollusks also got their start in the Cambrian period, and their shells also made good, recognizable fossils. There were likely myriad kinds of soft-bodied worms and other shell-less invertebrates wriggling around in the Cambrian mud, but their bodies almost always rotted away before they could form an impression in sediment that would eventually turn to stone, so we know little about them.

One of the most important Cambrian fossil beds in the world is the Burgess Shale, located in the northern Rockies of British Columbia. The Burgess Shale contains some of best preserved Cambrian invertebrate fossil specimens ever discovered. Some rather bizarre-looking creatures wandered the swamps and shallow seas of British Columbia back in the Cambrian. Their assortment of weird appendages, eyes perched on long stocks, and peculiar body shapes call to mind creatures from the imagination of a science fiction writer. Such creatures have to be seen to be believed, so I recommend a trip to the Walcott Quarry, where tour guides will show you the fossil beds and explain their significance. The quarry is in Yoho National Park near the town of Field, which lies between Lake Louise, Alberta, and Golden, British Columbia.

The Rocky Mountain region abounds in fossil sites containing early marine fossils, dinosaur fossils, freshwater fossils, and terrestrial fossils. Many of these fossil deposits are exposed in road cuts, stream banks, cliffs, and canyon walls. Some of these fossil beds can be seen quite easily from the side of the road, such as the dinosaur tracks visible near the town of Morrison, Colorado. Many of the best fossil beds are preserved in national, state, and provincial parks throughout the Rockies. Others require more detailed knowledge of local geology to find, but this information is available in regional geological guidebooks. Fossil hounds coming to the Rockies might expect to find horizontal layers of fossil-bearing rocks, stacked up like a layer cake of ancient life. This is seldom the case. The fossil beds are most often tilted—sometimes very steeply tilted—in this part of the world. Those mountain-building forces kept pushing up this way and pulling down that way, making life more complicated for the paleontologists who would eventually scramble over the hillsides.

Dinosaur Fossils

Dinosaur fossils attract quite a bit of attention today. This is not a new phenomenon but rather has been the case since the science of paleontology began in earnest during the mid-1800s. When people discover fossilized dinosaur bones in the ground, they generally find the time and money to get them dug up and shipped off to a museum. Accordingly, there are very few places where you can go to see dinosaur bones that have been left in the ground. One of the best of such places that does remain is Dinosaur National Monument, straddling the northern Colorado–Utah border. The largest quarry of Jurassic dinosaur bones ever discovered is on this site 20 miles east of Vernal, Utah. Paleontologist Earl Douglas discovered thousands of dinosaur bones at this site, including several nearly complete skeletons. A quarry was dug, and the site was designated a national monument in 1915. A visitor center has been built over the quarry to protect the fossilized dinosaur bones and skeletons. More than 2,000 dinosaur bones are exposed in the sandstone wall. This is the most productive Jurassic period dinosaur quarry in the world, providing more complete skeletons, skulls, and bones of dinosaurs than any other locality.

The Story of Lake Florissant

One of the most fascinating fossil beds in the Rockies is at Florissant, Colorado, where a series of volcanic eruptions sent alternating mudflows and ash deposits cascading down on the landscape. About 35 million years ago, a lake formed there when a lava flow dammed a stream valley. The lake was 12 miles (19 km) long and 2 miles (3 km) wide. A series of volcanic eruptions in mountains south of the lake laid layer on layer of ash that settled in the lake waters, preserving plants, insects, and fish skeletons. Ancient sequoia trees growing near the lake were buried by volcanic mudflows, and mineral-rich groundwater percolated through these deposits, replacing the organic cells of the wood and bark with replacement minerals in the process known as petrification (literally, "turning to stone"). Finally, the whole basin was buried by a massive mudflow that capped the lake sediments and allowed them to turn to stone before the process of erosion could wear them away. The resistant layer of stone that had capped the lake sediments and petrified trees was exposed to erosion, and the mudstones and shales of the ancient lake were once again exposed to the light of day. Since its discovery in 1874, the Florissant fossil beds have yielded thousands of fossil plants, insects, fish, and the bones of some primitive mammals.

Lake Florissant dominated the ancient forested valley. Surrounding the lake were lush ferns and shrubs, growing under towering redwoods, cedars, pines, and a colorful mixed hardwood forest of maples, hickories, and oaks. The climate was warm and humid, with mild winters. The forest was home to thousands of insects, and the lake was full of fish and mollusks. Birds and early forms of mammals inhabited the lakeshore. But the life in this valley came to an abrupt and violent end when a nearby volcano erupted. Volcanic mudflows buried the forest and blocked the lake drainage. The exploding volcano showered the countryside with millions of tons of ash, dust, and pumice. Caught in this deadly cloud, the insects, plants, and fish died; many fell to the lake bottom, where they were buried. These eruptions occurred again and again for perhaps as long as 5,000 to 10,000 years. In this way, fragments of life were trapped layers of volcanic sediments at the bottom of the lake.

About twenty million years before the Florissant fossil beds were being laid down in central Colorado, the Green River Formation fossil beds were forming in southwestern Wyoming. The Green River shales are fifty to fifty-five million years old and were laid down in a large, shallow lake. The lake eventually filled with sediments and chemical precipitates. The Green River beds are famous for their fossil fish, preserved in exquisite detail in the silty lake sediments. Other fossil organisms in the Green River shales include lizards, crocodiles, turtles, snakes, birds, and insects. Examples of these fossils, and the fossil beds themselves, can be seen at Fossil Butte National Monument, near Kemmerer, Wyoming.

Fossils from the Ice Age

Quaternary fossils are often not mineralized replacements of bones or plant parts but the actual remains of the animals and plants that have been preserved in unusual circumstances. They are unusual because the usual thing that happens to plants and animals after they die is that they decompose. However, the physical environments of the Rocky Mountain region allowed the remains of some ice-age flora

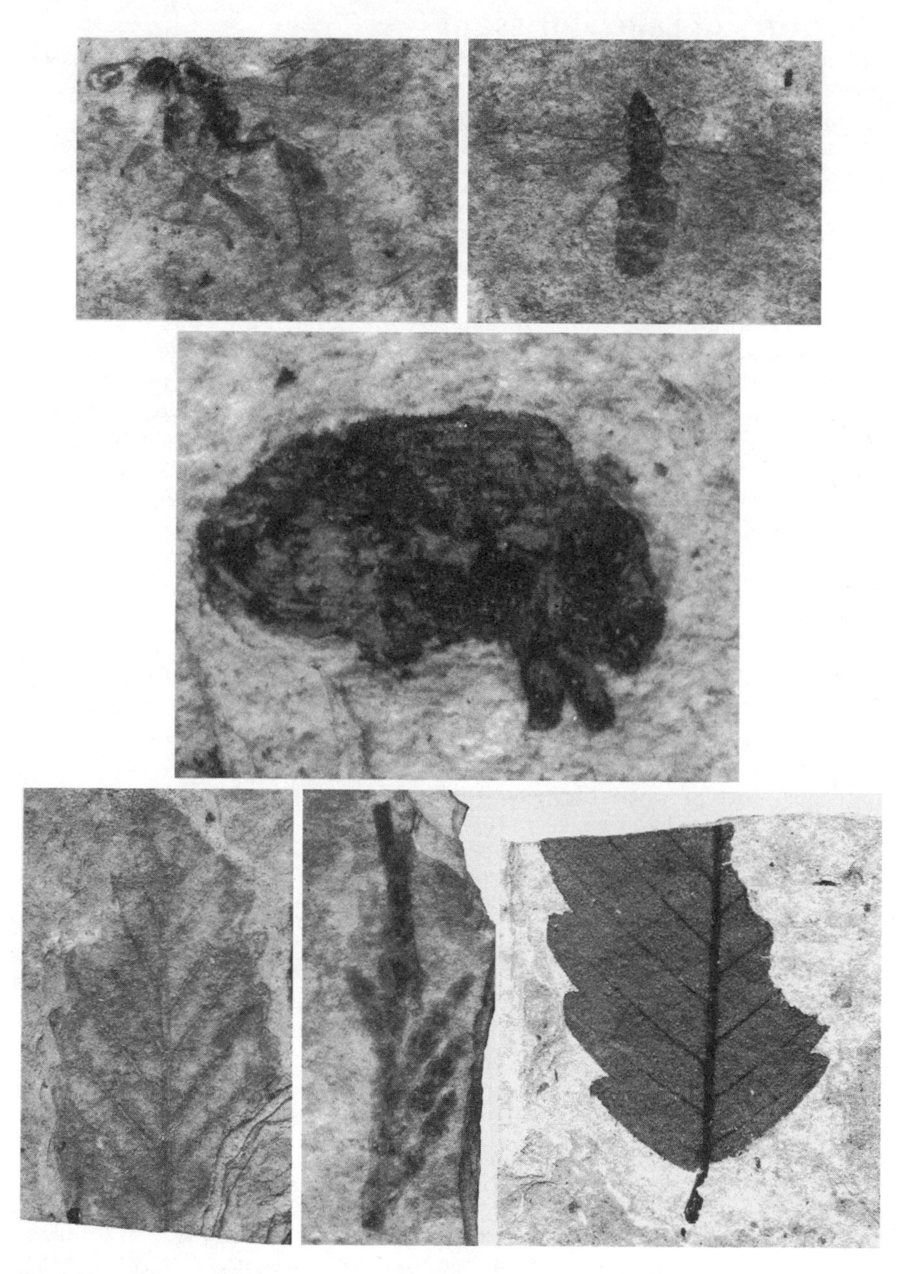

Thirty-five-million-year-old fossil insects and plants from Florissant, Colorado

and fauna to escape decomposition in a couple of ways. First, lakes and peat bogs sometimes accumulate plant and animal bodies in such a way that decomposition is greatly retarded because oxygen is shut out soon after the remains enter the water. If sediment accumulates quickly enough, it will keep oxygen away from the organic matter it contains. When these oxygen-starved layers build up for centuries and millennia, thick layers of peat and lake sediments can hold on to their fossil remains and preserve them for great lengths of time.

A second way for decomposition to be retarded is for the bodies of plants and animals to become dried out. Dry caves in the Rockies have served as mausoleums for the bones of ice-age animals that died there. One such cave is Natural Trap Cave, east of Yellowstone National Park near the Montana–Wyoming border. Natural Trap Cave contains bones ranging in age from 21,000 to 11,000 BP, from the last glaciation. The cave fauna include many extinct animals, including dire wolf *(Canis dirus),* short-faced bear *(Arctodus simus),* American lion *(Panthera leo atrox),* American cheetah *(Acinonyx trumani),* mammoth *(Mammuthus),* four kinds of extinct North American horses *(Equus),* American camel *(Camelops),* woodland musk-ox *(Symbos cavifrons),* and extinct species of bighorn sheep *(Ovis),* bison *(Bison),* and pine marten *(Martes).* The cave fauna also are comprised of species that are no longer native to this region, including arctic inhabitants such as the collared lemming *(Dicrostonyx hudsonicus)* and the arctic hare *(Lepus arcticus).* In addition to these exotic elements, Natural Trap Cave preserves the remains of many mammals still native to northwestern Wyoming, including antelope *(Antilocapra americana),* gray wolf *(Canis lupus),* cottontail rabbit *(Silvilagus),* chipmunk *(Tamias),* pocket gopher *(Geomys),* and several species of rodents. Most of the extinct species from these cave deposits are large mammals, whereas all of the species in the fossil assemblages that are still living today (whether in Wyoming or elsewhere) are small- to medium-sized animals. There are many grazing animals in the fossil assemblages. The region east of Yellowstone was likely grassland throughout the late Pleistocene, just as it is today.

However, the variety of animal life seems to have been far richer then. Imagine the fauna of the African savannah transported to a cooler climate. In Pinedale times, this image would have fit the Yellowstone region reasonably well. In place of African elephants, there were Columbian mammoths *(Mammuthus columbi).* In place of African cheetahs and lions, there were North American cheetahs and lions. Short-faced bears, gray wolves, and dire wolves hunted camels, musk-oxen, and American horses, as well as antelope and bison. By 11,000 BP, many of these animals no longer existed. Their Old World relatives managed to survive through the Holocene, but in the New World they became extinct.

Yet the comparisons with the modern fauna of Africa may be misleading, because the glacial climate of northern Wyoming was much different from that of the African savannah. There is a strong arctic–subarctic element in the ice-age fauna of Wyoming. In addition to collared lemming and arctic hare, musk-oxen and caribou *(Rangifer tarandus)* also ranged across the grasslands of Wyoming during the late Pleistocene. These species are found only in Alaska and northern Canada today. So although the plains of Wyoming may have been as dry as the modern African savannah, they were certainly far colder.

The extinction of large mammals in Wyoming at the end of the Pleistocene was part of a continental-scale disaster. All the proboscideans (mammoths and mastodons), the horses, camels, giant sloths *(Nothrotheriops),* and many other species became extinct in North America at the end of the last glaciation. Why did this happen? The obvious answer might seem to be that these cold-adapted animals could not tolerate the warm climates of the Holocene. This argument might be convincing if it were not for the fact that the same cold-adapted species and their ancestors had withstood the warm climates of about a dozen previous interglacial periods, at least one of which was probably substantially warmer than anything yet experienced in the Holocene.

So climatic warming per se was not enough to extinguish the large mammals. Some unique biological or environmental factors must have influenced large mammal populations at the beginning of the Holocene. Some argue that human beings were the most important agent in dispatching the North American megafauna. Paul Martin, geologist at the University of Arizona, coined the phrase "Pleistocene overkill" to describe how people may have hunted these animals to extinction. The theory suggests that the megafauna of North America were especially vulnerable to late Pleistocene (Paleoindian) hunters because the people were newcomers on the North American continent at that time, and the animals, unaccustomed to human hunters, had little natural fear of them. The theory holds that this new hunting pressure, combined with rapid climate change, wiped out most of the megafauna on this continent. We will probably never know if the overkill theory is the right one, because the fossil evidence is so spotty. Whatever the cause of the extinction, we are left with a rather poor collection of large animals that have made it through the Holocene.

Another site, not too far from Natural Trap Cave, offers a fascinating glimpse into the life of ice-age mammoths. In the Black Hills of South Dakota (technically outside the scope of this book, but close enough to the Rockies to be relevant to our discussion), limestone deposits were dissolved by groundwater in the late Pleistocene, creating large sinkholes in the landscape. At one such locality at the present town site of Hot Springs, mammoths became trapped in the sinkholes, and their bones were preserved as the holes filled in with sediment. The Mammoth site at Hot Springs is now preserved as a museum, displaying some mammoth bones that have been excavated and others that have been left in place for visitors to see.

During the past two million years, glaciations, or ice ages, have come and gone repeatedly in the Rockies. As I discussed previously, over the geological time span that represents the history of the current species of flora and fauna, cold climates have been far more prevalent than warm climates. This is a hard concept for us to grasp, because human civilization has grown up entirely within the current interglacial period, the Holocene. We know nothing else. Our ancient ancestors, those who hunted woolly mammoth and sought shelter in caves, have left us no written record of what it was like to live in glacial times.

During glacial intervals, the life zones of the Rockies were pushed downslope. The degree to which these life zones shifted depended on the severity of glacial cold; we know from the geological record that not all glaciations were alike. We know most about the last glaciation, because no subsequent glaciations have come along to smear its paleontological and geological record.

During the last glaciation, the boundary between alpine tundra and subalpine forest (upper treeline) pushed down the mountain slopes by hundreds of meters, and coniferous forest crept out onto the plains in some places. At the end of the last glaciation, the various groups of biological communities moved back north and back upslope to their present elevational zones. However, these postglacial shifts did not come about through the movements of whole communities. As with any major environmental change, each species responded individually, and the composition of the various communities changed, sometimes quite dramatically. Rather than an orderly transition up slopes or from south to north, the biological shifts that took place at the end of the ice age were far more chaotic.

Given that ice ages are the norm and interglacial warmth is the exception, it follows that the cold-hardy biota of the mountains and high latitudes occupied more real estate through most of the Quaternary period than they do today. Conversely, the warmth-loving biota are now enjoying a brief respite from the severe restrictions placed on them by those seemingly interminable ice ages. The shift from glacial to interglacial climates took place perhaps seventeen times in the Pleistocene. Because of the nature of the fossil record (younger glaciations tend to wipe out the evidence of previous events), our best evidence concerns the transition from the Pinedale Glaciation to the Holocene, about 10,000 years ago. As with all large-scale environmental transformations, this last one wrought havoc with the existing biological communities. We explore these changes and what they mean for modern ecosystems, working from north to south.

The Canadian Rockies were the last part of the Rocky Mountain chain to emerge from the ice at the end of the last glaciation. Although parts of the southern Rockies were deglaciated by about 14,000 years ago, the northern Rockies remained glaciated until about 10,000 to 11,000 years ago. For instance, the Lake O'Hara region in Yoho National Park, British Columbia, and Crowfoot and Bow Lakes in the Alberta Rockies were all free of ice by 10,100 BP. A little farther south at Marias Pass, Montana, the first evidence of deglaciation comes at 12,000 BP, and by 11,200 BP the ice had retreated to local mountain valleys in Glacier National Park. By 11,000 to 10,000 BP ice had withdrawn to the same high-mountain cirques and shaded niches where it remains today.

According to paleobotanist Mel Reasoner, the first vegetation to invade the high country of the Canadian Rockies after the ice retreated was a mixture of shrubs and herbs, dominated by sage, grasses, and alder shrubs. This plant cover remained dominant in the Lake O'Hara region until after about 8500 BP, when the first conifers became established. The shrub-herb vegetation that colonized the bare mineral soils left by retreating glaciers began the slow process of stabilizing the landscape and building up organic matter in the soils. Summer temperatures may already have been quite warm as these pioneer communities became established, but trees could not invade until the organic content and nutrients in the soils had been developed under the cover of herbs and shrubs. The first forest to make its way up to the high country was composed of whitebark pine *(Pinus albicaulis)* or limber pine *(Pinus flexilis),* fir *(Abies),* spruce *(Picea),* and lodgepole pine *(Pinus contorta).* Lower down the valleys, there is fossil evidence that lodgepole pine was already established before 11,000 BP. Presumably the other tree species were likewise established at these lower elevations, but they could not colonize the high country until the bare mineral substrates were transformed into the kind of soils that nurture conifer seedlings. Spruce was likewise residing in lower montane and foothills localities of the northern part of the Canadian Rockies at the beginning of the Holocene, whereas pine was the dominant conifer group in the southern foothills.

During an interval of warmer than modern climate in the early Holocene, conifers grew at higher elevations in the northern Rockies than they do today, as evidenced by the remains of logs and other macrofossils in bogs and ponds that are above the modern treeline. In mid-Holocene times, spruce became dominant at high-elevation sites, at the expense of limber or white pine. Forests in the Crowfoot Lake region also contained lodgepole pine during this interval (about 4200 to 900 BP). Mountain glaciers advanced in the northern Rockies during the past thousand years, in what geologists call the "Little Ice Age." Forest-tree density decreased throughout the highlands of the northern Rockies, and open ground covered with alpine tundra expanded downslope to elevations below modern treeline, which has only been established at most in the past few hundred years. Similar Little Ice Age events took place throughout the Rockies, though with decreasing intensity of effects toward the south.

The late Pinedale biota of Montana had two sets of glacial ice to contend with. The mountains had montane glaciers that spilled out onto valley floors, and the lowlands of the northern sector of the state were buried in continental ice sheets. An ecosystem resembling arctic tundra bordered the southern edge of the ice in Montana. It was essentially a cold prairie, with a mixture of arctic and grassland species. Southwestern Montana was home to mammoths, musk-oxen, and caribou. On the other hand, this region also supported populations of prairie dogs *(Cynomys ludovicianus),* sage voles *(Lemmiscus curtatus),* and pocket gophers. None of the latter group of animals is adapted to arctic conditions, so the climate, although decidedly colder than today, was not an arctic climate. Winter snows persisted on the ground longer than they do now, and the summers were cool.

Fossil pollen in lake basins in northern Montana provide clues that help reconstruct the postglacial vegetation history of the region. To the east of Glacier National Park, palynologist Cathy Whitlock has studied the transition from late glacial to Holocene environments from pollen in the sediments in Guardipee Lake and the history of the Holocene prairies at Lost Lake. West of the park, Richard Mack and colleagues have analyzed postglacial pollen from Tepee Lake and McKillop Creek Pond in the Kootenai Valley. At Marias Pass, just south of the park, Susan Short has studied pollen in sediments from a late glacial pond.

The Guardipee Lake basin was covered by the Two Medicine Glacier during late Pinedale times. The glacier retreated and sediments started accumulating just before ash from an eruption of Mt. St. Helens was deposited in the lake, about 12,000 BP. The pollen from this lake provides a record of regional vegetation that begins just after deglaciation. As in the Canadian Rockies, the vegetation cover that developed on the newly exposed landscape was dominated by grasses with some sagebrush. Nearby mountain slopes to the west were already supporting pines, spruce, and fir. Closer to the lake, junipers *(Juniperus)* were becoming established. Juniper apparently does not compete well with other conifers in regions where they are all capable of growing, but in the early postglacial times tree species were only just beginning to disperse across vast deglaciated landscapes, and junipers were one of the first conifer groups to take advantage of these landscapes. Willow *(Salix)* and aspen *(Populus tremuloides)* were probably also growing in local wet areas during deglaciation. Above about 4,920 feet (1,500 m) elevation, tundra, fellfields, and the remaining ice were holding on.

By 11,500 BP, the vegetation surrounding Guardipee Lake reflects a change in climate. An increase in aridity was inferred from a rise in sagebrush and other dry-land plant pollen, at the expense of grasses and other herbs. In addition, plants adapted to brackish water started growing in Guardipee Lake. This finding suggests that the lake level was dropping. The lake dried out after 9300 BP and then went through a series of wet and dry episodes for the rest of the Holocene.

Meanwhile, in the Kootenai River Valley, the sediments in Tepee Lake began accumulating and collecting pollen. The Cordilleran Ice Sheet withdrew from the site before 11,200 BP. By this time, regional climate was clearly quite warm, because the newly deglaciated terrain was immediately colonized by pines (probably whitebark pine, based on the shape and size of the pollen grains). Smaller numbers of spruces were also present, with a groundcover of grasses. By about 11,000 BP, conifer forest was growing right up to the edge of the waning Cordilleran Ice Sheet.

From 11,000 to 7000 BP, regional forests were dominated by pines, Douglas-fir *(Pseudotsuga menziessii),* and larch *(Larix)*. Open ground regions, such as meadows and dry hillsides, were covered with sagebrush. Based on the pollen evidence, the climate during this interval was warmer than the climate during the Pinedale deglaciation.

In the Kootenai River Valley, drier conditions prevailed after 7000 BP, and dry-adapted conifers (ponderosa pine, *Pinus ponderosa,* and lodgepole pine) became regionally important. This aridity was especially strong at about 6700 BP, when sagebrush replaced conifers. After 4000 BP, moister conditions returned, accompanied by Engelmann spruce *(Picea engelmannii)*. Modern climatic conditions and vegetation became established about 2700 BP, heralded by the establishment of western hemlock *(Tsuga heterophylla)*, a tree that characterizes modern forests from western Washington to northwestern Montana.

On the plains east of the park, the pollen record from sediments in Lost Lake records changes in prairie vegetation through the Holocene. From 9400 to about 6000 BP, this region was an arid grassland with a drier climate than is seen today. Following 6000 BP, cooler, moister conditions allowed a more mesic (medium-moisture conditions) grassland to develop on the plains, shrubs increased in wetland regions, and conifer forest expanded in nearby mountains.

Based on the evidence from the Kootenai River Valley, the modern ecosystems of northwestern Montana did not become established in their current form until 8,000 years after Pinedale ice retreated from this region.

In Wyoming, some of the best records of late Pleistocene flora and fauna come from the Yellowstone region. Yellowstone National Park was buried in a glacial ice sheet during the last glaciation, and the effects of that glaciation can be seen throughout the park and adjacent areas.

The park region was covered by Pinedale ice from about 30,000 to 14,000 BP. Several late Pinedale-age sites in and around the park have provided pollen assemblages. These have been studied chiefly by Richard Baker and Whitlock. Blacktail Pond lies near the northern park boundary. This site was deglaciated about 14,500 BP, and the pond sediments began accumulating pollen about 14,000 BP. The oldest pollen assemblages indicate tundra vegetation, giving way to spruce parkland by about 12,800 BP. The northern sector of the park is currently much drier than the Central Plateau and southern highlands. This has apparently held true throughout much of the late Pleistocene, if not before, because precipitation patterns are strongly tied to regional topography. So the tundra vegetation at Blacktail and Slough Creek ponds, farther east, lacked the moisture-loving plants, such as bog birch *(Betula pumila)* and some alpine species, seen in ancient vegetation records from sites such as Buckbean Fen, farther south in the park. In addition, late glacial forests farther south included fir and poplar, whereas the Blacktail Pond region supported only spruce parkland in the early part of the late glacial warming. The differences in climate and vegetation between north and south became more pronounced during the Holocene.

Starting about 12,000 BP, the Yellowstone region became forested. Engelmann spruce was the first conifer to become established in most parts of the park, followed by whitebark pine and lodgepole pine. By about 9500 BP, lodgepole pine had become regionally important. A combination of increasing temperature and precipitation fostered the establishment of pines from 11,800 to 9500 BP. The Central Plateau region of Yellowstone is underlain by rhyolite bedrock. The soils that develop from this parent material are relatively infertile, and lodgepole pine is the only regional tree species that does well on this type of soil. Because of these circumstances, the Central Plateau region remained treeless until about 10,000 BP, when lodgepole pine invaded the region.

During the past 10,000 years, changing climates in the Yellowstone region brought about some large-scale changes in regional vegetation. Today the northern part of the park is considerably drier and warmer than the highlands to the south. This is easily appreciated in late spring, when the southern parts of the park are buried under meters of snow and the Mammoth region is often snow free—the "banana belt" of Yellowstone. Thanks in large part to the work of Whitlock, we have come to understand how topographic differences between these regions affected Holocene environments. During the early Holocene (9,500 to 7000 BP), the northern part of the park experienced wetter climates than it does today, whereas the southern region was warmer and drier. Paleoclimate researchers believe that these changes were brought about by shifts in atmospheric circulation, linked with shifts in precipitation patterns. Paleoclimate reconstructions suggest that the increased summer insolation during the early Holocene created summer weather patterns dominated by high pressure in the southern Yellowstone region. Relatively warm and dry conditions tend to increase fire frequency in this region, and increased numbers of forest fires helped to maintain such fire-adapted species as lodgepole pine, Douglas-fir, and aspen. These trees outcompete other species in fire-prone regions. For instance, the cones of lodgepole pine are triggered to release their seeds when they are heated by forest fires. This characteristic ensures that a large crop of lodgepole pine seedlings will sprout in recently burned landscapes, overwhelming the seedlings of other trees. Thus the southern Yellowstone region supported forests dominated by lodgepole pine and Douglas-fir from 9500 to 5000 BP.

The same atmospheric conditions that fostered a warm and dry climate in southern Yellowstone during the early Holocene brought increased moisture to the northern sector. This region, plus parts of central and eastern Wyoming, apparently received increased precipitation from summer monsoons. From 9500 to 7000 BP, the northern Yellowstone region supported forests of lodgepole pine, juniper, and birch. The region probably looked quite different than it does today; we see broad parklands of grasses and sagebrush with conifers growing mostly on moister hillsides. This pattern broke down about 7,000 years ago, as the northern region became drier and the vegetation opened up into Douglas-fir parkland. This increasing aridity culminated in the modern Douglas-fir–lodgepole pine parkland ecosystem by about 1600 BP. By 5000 BP, the southern region began receiving increased moisture and the modern closed spruce–fir–pine forest became established.

Paleoenvironmental reconstructions from the southern Rockies are more sketchy than those developed for the Yellowstone region. The oldest Pinedale site in the region is near the Mary Jane ski area, at Winter Park, Colorado. Here pollen in lake sediments laid down during an interstadial interval before the last major Pinedale ice advance (circa 30,000 BP) records a sequence of vegetation beginning with open spruce-fir forest with herbs and shrubs adjacent to the lake. This is followed by a colder phase, in which alpine tundra replaced the subalpine forest. The youngest (uppermost) sediments in this lake bed reflect the return of spruce forest to the vicinity before the advance of late Pinedale ice. The Mary Jane site is at an elevation of 9,450 feet (2,882 m), in the lower part of the modern subalpine forest. The existence of alpine tundra at the site in mid-Pinedale times translates into a depression of treeline by more than 1,640 feet (500 m).

The next indication of late Pinedale environments in the region comes from the oldest lake sediments from Glacial Lake Devlin (22,400 BP), a subalpine site in the Colorado Front Range. Geologists Thomas Legg and Baker found pollen evidence for a treeless landscape at the site. The pollen diagram is dominated by sagebrush. Scattered pollen grains of tundra plants (which are not abundant pollen producers) and low numbers of conifer pollen grains suggest that the site was above treeline in the late Pinedale. The pollen assemblages from this site are in many ways similar to those described by Baker and Whitlock from the earliest postglacial environments in Grand Teton and Yellowstone National Parks. In both instances, mixtures of sagebrush and alpine tundra plants suggest steppe-tundra. This vegetation probably developed in the sort of cold, dry climate seen today in parts of Siberia.

At the Mary Jane site, peat layers were again deposited after the retreat of late Pinedale ice. Pollen samples that range in age from 13,740 to 12,700 BP show that the site was occupied by alpine tundra. After 12,600 BP willow shrubs expanded across the bog, and spruce trees were probably growing adjacent to the site, after an absence of about 18,000 years.

The vegetation history of the Rocky Mountains of northern Colorado has been interpreted from fossil pollen data. One of the study sites is Long Lake, in the Indian Peaks Wilderness. The lake lies near treeline in the upper subalpine zone. At this site, pollen from the lake sediments shows that alpine tundra vegetation became established after deglaciation and persisted from 12,000 BP to 10,500 BP, when spruce and fir arrived. However, the presence of these conifers does not mean that subalpine forest became established at that time, because elements of the alpine tundra persisted there until about 9500 BP. This trend is also seen from data collected at La Poudre Pass, just north of Rocky Mountain National Park, where a mixture of alpine tundra and spruce woodland developed after deglaciation, from 9800 to 9100 BP. Sagebrush, grasses, and other herbs dominated the ground cover. From about 9000 to 6800 BP, spruce-fir forest and some pine (probably lodgepole pine) grew at La Poudre Pass and Long Lake, but treeline was still probably below its modern elevation at the former site. The vicinity of the bog at La Poudre Pass still had much open ground with herbaceous cover, but alpine tundra vegetation had retreated upslope by this time.

The time of maximum Holocene warmth, as indicated by pollen assemblages from La Poudre Pass, Long Lake, and other sites, came between 6500 and 3500 BP. During this interval, pine expanded upslope into the modern subalpine zone throughout the Front Range. The upper limit of spruce-fir forest may have crept higher during this interval as well, but the evidence for this conclusion is unclear at the sites. During the past 3,500 years, pine retreated back to the elevations of the modern montane forests, and levels of herbaceous pollen increased once again. During this interval, the modern pattern of regional vegetation became established.

Selected References

Alt, D., and D. W. Hyndman. 1986. *Roadside Geology of Montana.* Missoula, Mont.: Mountain Press.

———. 1989. *Roadside Geology of Idaho.* Missoula, Mont.: Mountain Press.

Briggs, D. E., D. H. Erwin, and F. J. Collier. 1994. *The Fossils of the Burgess Shale.* Washington, D.C.: Smithsonian Institution Press.

Chronic, H. 1980. *Roadside Geology of Colorado.* Missoula, Mont.: Mountain Press.

———. 1987. *Roadside Geology of New Mexico.* Missoula, MT: Mountain Press.

Elias, S. A. 1996. *The Ice-Age History of National Parks in the Rocky Mountains.* Washington, D.C.: Smithsonian Institution Press.

Gould, S. J. 1989. *Wonderful Life: The Burgess Shale and the Nature of History.* New York: W. W. Norton.

4

Plant Life

The Rocky Mountains are home to more than 500 species of plants, displaying a diversity that results from the wide variety of habitats found in the mountains. The range of elevations in the Rockies provides climatic conditions that in other regions span half the continent. In other words, to find the same level of climatic variability that can be found on one slope of a tall mountain peak, you would have to travel from the semidesert grasslands of New Mexico to the arctic tundra of Canada. The wide variety of rock outcrops found in some Rocky Mountain ranges provides for great variety of soil types, another crucial element in plant habitats. Also, the complex precipitation patterns that form around the different slopes of the mountains create striking differences in soil moisture. In some places, extremely dry slopes sit right next to extremely wet ones. Of course none of these habitat elements stands alone. They all interact to form complex habitat patterns. Keep in mind that the sweeping generalizations that follow are full of exceptions, local variations, and unmentioned complexities. So strap on your hiking boots, and let's take a summer stroll up a typical mountainside in the Rockies.

The basic pattern of plant life is fairly consistent throughout much of the Rockies. In the foothills zone, where the mountains meet the plains, the slight rise in elevation often wrings a little bit of extra moisture out of the clouds. On drier (often south-facing) slopes, the foothills zone is usually dominated by the grasses and other herbs associated with the nearby plains. On the moister (often north-facing) slopes, shrubs and some conifers can grow. The eastern foothills of the Rockies share many plants in common with the Great Plains, whereas the foothills of the southern Rockies in southern Colorado and New Mexico share plant species with the arid southwest. On the western slope of the southern and central Rockies, the foothills flora have much in common with that found in the Great Basin. The foothills on the western slope of the northern Rockies have plant species in common with the Pacific Northwest.

The reasons why the foothills of the Rockies share so many plant species with adjacent lowlands are mainly historical. Since the end of the last ice age, changing cli-

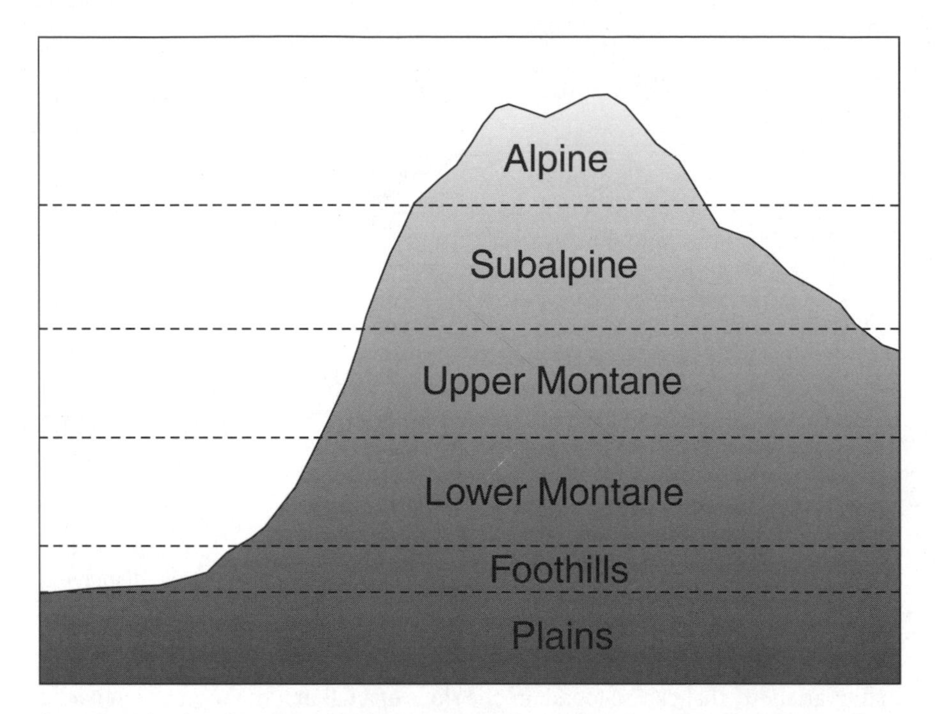

Generalized drawing of a mountain landscape in the Rockies, showing the main ecological zones

mates have driven plant species to migrate and become established in new regions. The mixture of plants that we see today in any given region is the result of the latest round of shuffling in this process. A thousand-year cold climate may force a warm-adapted species to die out in one region, whereas 2,000 years of warm climate may allow that species to expand into new territory. As climate warmed and cooled, the various plant species moved up the mountainsides, down into valleys, out onto the plains, and back again. It is hard for us to perceive these climatic forces at work, because they operate on scales much longer than a human lifetime.

As we move up into the mountains, we will leave the grasslands of the plains behind. Except for some grassy meadows, you are going to be seeing forests of various shapes and sizes until we get to the alpine tundra. The lowest of these forest zones is called the montane zone. Montane habitats are consistently cooler and wetter than the drier habitats of the foothills, so even the south-facing slopes support trees. Montane forests are dominated by pines and other conifers. Depending on the region, these forests range from dense stands of trees to more open stands, with large patches of herbs and shrubs between trees (parkland).

Now it is time to hike up to the highlands. You've probably noticed that the air is getting thinner and cooler. The forests are starting to look different, too. The higher elevation forests of the Rockies are lumped together into what is called the subalpine zone. Subalpine forests are often dominated by spruce and fir trees. Subalpine winters are longer and summers shorter than montane winters. The snowpack is deeper and it melts out later in the year, creating a shorter growing season. It is not surprising, then, that the subalpine forests of the Rockies have a lot of features in common with the subarctic forests at the northern edge of the boreal zone in Canada and Alaska.

The last leg of our imaginary trip up the mountain takes us to the alpine zone, the land above the trees. Why don't trees grow here? The short answer is that it is just too cold. Treeline (the upper limit of trees) tends to occur at or near the elevation where the average July temperature is 50°F (10°C). Above this elevation, the growing season is too short and the summers too cold to allow tree seedlings to get established.

Much of the land above the trees is bare rock. Some regions support alpine glaciers, but the unglaciated landscapes that have soils are clothed in alpine tundra. The region of the world with closest affinities to the alpine tundra is the arctic tundra of northern Alaska and Canada. The arctic tundra is essentially the "mainland" of the tundra habitat. The mountaintops along the Rockies are thus habitat "islands" of alpine tundra, surrounded by a "sea" of conifer forests. The alpine flora is highly adapted to life in a cold, dry climate. It grows close to the ground (the better to keep out of chilling, desiccating winds), and it grows slowly. Plants that have to set seed every year (annuals) do not fare very well in the alpine tundra, because many years may not offer summer conditions that are warm enough for long enough to allow these plants to complete their life cycle. Perennials are by far the dominant group of plants. These species can survive year after year, slowly storing up reserves of chemical energy in their root systems that eventually allow them to produce a good crop of seeds.

Though upper treeline is thought to be controlled chiefly by summer temperatures, lower treeline, the elevation at which lower montane forest gives way to prairie vegetation, is thought to be controlled chiefly by precipitation. In other words, it is the lack of moisture that keeps conifer species from invading the prairies. Twentieth-century towns that grew up and down the flanks of the Rockies have demonstrated this principle quite well. City dwellers have been growing montane forest tree species in their yards with great success for more than a century, simply by enhancing the natural precipitation with a little irrigation.

Although the Pleistocene glaciations brought substantially colder temperatures to the Rockies, geologists are still debating whether the glacial intervals were substantially wetter than today. In glacier modeling, it is possible to grow glaciers by either cooling the temperature or adding precipitation or with a combination of the two. Because this issue remains unresolved, we are not at all certain that montane forests extended out onto the plains for any great distance. The fossil plant record from the Great Plains immediately east of the Rockies does not indicate that conifer forests invaded there during the last glaciation. It seems most likely that the boundary between montane forest and prairie was pretty much the same as it is now, at least during the past 100,000 years.

The various zones of vegetation in the Rockies also provide habitats for the animal life of the region. Some animals move up and down freely between zones, but a surprising number of animals (especially small mammals, reptiles, amphibians, and invertebrates) inhabit only narrow elevational zones. The trees of the mountain forests do a lot to shape the habitat of other creatures. Just by being there, forests reduce the amount of sunlight reaching the surface. Trees also slow the winds that come whistling through the mountains, so that the wind speeds at the surface may be considerably slower than the wind speeds above the tree tops. Conifers drop needles, cones, and twigs that accumulate in considerable thicknesses below stands of trees. This layer, called "duff," is mostly made up of needles, and water percolating through the duff layer becomes more acidic, preventing many kinds of plant seedlings from becoming established beneath the trees. The roots then fan out from the base of the trees and hold onto large quantities of soil, keeping erosion at bay.

Important Plant Communities

Keeping in mind my cautions about generalities, let's proceed with regional synopses of important plant communities. My treatment of this subject will progress southward from the northern margin of the northern Rockies to the southern edge of the southern Rockies. Because the communities on the eastern slope of the Continental Divide tend to be different from those on the western slope, these two regions will be covered separately throughout the rest of this chapter.

Plant Communities of the Northern Rockies

The northern Rockies comprise a topographically diverse region that ranges from about 46° N to about 54° N latitude. At one extreme, the Canadian Rockies of Alberta and southeastern British Columbia form an almost uninterrupted arc of highlands, more than 250 miles (400 km) long. At the other extreme, the Rockies of Montana and Idaho are broken up into dozens of small ranges; these ranges occur in clumps in western Montana and southern and eastern Idaho. One of the intriguing aspects of all of the northern Rockies is that they are not very tall. Even a quick glance at a topographic map of this region reveals that there is nothing above 12,000 feet (3,660 m) in the entire region. On the other hand, these mountains are far enough north that even elevations of 9,000 feet (2,750 m) are mostly above treeline. Also, the vertical relief of these mountains above adjacent lowlands tends to be at least as great if not greater than elsewhere in the Rockies. For instance, in the Colorado Front Range, the neighboring plains average about 5,000 to 5,500 feet (1,520 to 1,680 m) and the peaks rise to 13,000 to 14,000 feet (3,960 to 4,270 m), giving a vertical relief averaging about 8,200 feet (2,510 m). In the northern Rockies, the lowland regions average 1,000 to 3,000 feet (300 to 910 m), and the peaks rise to 11,000 to 12,000 feet (3,350 to 3,660 m), giving an average vertical relief of 9,500 feet (2,900 m).

These are the cold, wet Rockies. In the Canadian ranges, many peaks are capped with glaciers and snowfields, features that are rare or absent from mountaintops farther south along the Rocky Mountain chain. During the Pleistocene, the mountains of the northern Rockies did not just grow glaciers, they were *buried* in them. Only the tallest peaks stood above the regional ice sheets of this region. This is an important fact that affects the regional ecology. Plant communities were wiped out in this region during the last glaciation, and the whole region had to be recolonized after the glaciers melted, from the foothills to the mountaintops.

Since the Rockies trend from southeast to northwest, the farther north you go in the Rockies, the closer you get to the source of Pacific moisture. This is the dominant source of precipitation for the region, as storm tracks from the Gulf of Mexico and the Gulf of California tend to dissipate before they reach the latitudes of the northern Rockies. The western slopes of most of the ranges in this region receive substantial amounts of precipitation, and many of the forests on the western slope have a distinctive Pacific Northwest element in their composition. In contrast to this, the eastern slopes of the Northern Rockies are often in the rain shadow of their western-slope counterparts. The eastern-slope forests are therefore much drier, and their composition more closely resembles that of forests in the drier ranges of the central and southern Rockies.

Alberta Ranges

The montane zone of the eastern slope of the Continental Divide in Alberta is a region characterized by relatively dry climate. Banff receives an average of 18½ inches (470 mm) of precipitation per year, and Jasper receives only 16 inches (406 mm) per year. Winter weather oscillates between extremes. There are periods of intense cold, ushered in by arctic air masses moving south, but there are also periods of relatively warm conditions that develop when moist Pacific air masses move eastward over the continental divide. Banff receives an average of 30 inches (760 mm) more snow than Jasper; the increased snowpack often gives the Banff region a later start to the growing season but moister soils than those of the Jasper region, where soil moisture available to plants is often exhausted by the first of July.

Montane forests of Alberta are dominated by Douglas-fir *(Pseudotsuga menziessii)* and white spruce *(Picea glauca)* stands. The driest montane woodlands of southwestern Alberta are dominated by Rocky Mountain juniper *(Juniperus scopulorum)*. Douglas-fir stands are most common in the Porcupine Hills, and in the Bow, Kananaskis, and Oldman River Valleys southwest of Calgary. The Ponderosa pine forests that characterize the lower montane regions of the eastern slope of so much of the Rockies extend north only to about 47° latitude. The lower montane slopes in Alberta meet prairie dominated by fescue *(Festuca)* grasses as far north as Calgary. North of there, coniferous forests spread out onto the lowlands. The northern limit of Douglas-fir in Alberta occurs near Lake Brule, at about 53° N latitude. Montane forests in Jasper National Park range in elevation from about 3,000 to 5,000 feet (900 to 1,500 m).

Douglas-fir stands are found on warm, dry sites to mesic sites, such as occur on north-facing slopes of the foothills grassland region and on south- and west-facing slopes, well-drained river terraces, and rocky ridges in the mountains. In the latter localities, Douglas-fir is often accompanied by limber pine *(Pinus flexilis)* and whitebark pine *(Pinus albicaulis)*. The understory of Douglas-fir stands on north-facing slopes, and some subalpine spruce-fir forests are characterized by feather mosses. More open Douglas-fir woodlands have an understory dominated by grasses, and shrubs are more scarce in these stands.

The foothills of southern Alberta are dried by frequent chinook winds in winter. In addition to Douglas-fir stands, the silty soils of this region support lodgepole *(Pinus contorta)* and limber pines, aspen *(Populus tremuloides)*, balsam poplar *(Populus balsamifera)*, and a wide variety of grasses.

As elsewhere in the Rockies, aspen and lodgepole pine typically become established in Alberta's montane following disturbances, such as fires, landslides, or

Montane forest, Icefields Parkway, Alberta

avalanches. Aspen groves provide habitat for a lush understory of shrubs and herbs. Balsam poplar also grows in these disturbed sites.

Lodgepole pine becomes more important in the north, near the northern limit of Douglas-fir. Lodgepole pine is the dominant tree species of montane slopes and valleys in Jasper National Park. It also ranges upslope into the subalpine forest where fire has occurred. The other dominant montane forest tree of this region is white spruce. It is found on moist, cool sites, such as north-facing slopes, ravine bottoms, and young river terraces. Black spruce *(Picea mariana)* grows in wetter localities, such as lake margins, seeps, and peatlands.

Montane meadows develop on valley floors in the front ranges and foothills of Alberta. Chinook winds gobble up much of the winter snowpack in these openings, and most of the moisture comes as summer rain. These meadows are home to many species of wildflowers. These include daisy fleabane *(Erigeron compositus),* wild gaillardia *(Gaillardia aristata),* harebell *(Campanula rotundifolia),* and pasque flower *(Pulsatilla patens).*

The subalpine zone of Alberta is clothed in spruce-fir forest, chiefly Engelmann spruce *(Picea engelmannii)* and subalpine fir *(Abies bifolia).* These upper-elevation forests receive substantially more precipitation than the montane forests below. Subalpine meadows in the Canadian Rockies are a treat to the eyes. The vegetation can be quite lush, and depending on the time of year, the variety of wildflowers can be outstanding. Some of the common meadow species in this region are cow parsnip *(Heracleum lanatum),* fleabane, valerian *(Valeriana),* and ragwort *(Senecio).* Steep rocky slopes in the upper subalpine have thin soils and are exposed to almost continuous blasts of cold winds. Accordingly, these highlands support only prostrate trees, including Engelmann spruce, subalpine fir, common juniper *(Juniperus communis),* and limber or whitebark pine.

The ecotone (approximate boundary) between subalpine forest and alpine tundra in the Canadian Rockies occurs at elevations ranging from about 4,900 feet (1,500 m) in moist localities to about 7,000 feet (2,130 m) in dry localities. If you hike across the tundra, you will find three main types of vegetation: dry steppe meadows dominated by grasses, moist tundra with heather *(Cassiope),* cushion plants, dwarf willows and bog birch *(Betula glandulosa),* and wetlands (tundra ponds and fens) dominated by cotton grass *(Eriophorum callitrix).* Alpine wildflowers in the Canadian Rockies, as elsewhere in the alpine, are close to the ground, but it is worth the effort of getting down on hands and knees to take a close look at them. They are marvels in miniature, with startlingly bright colors and delicate forms. Among these little beauties are alpine forget-me-nots *(Eritrichum aretioides* and *Eritrichum nanum),* alpine anemones *(Antennaria parvifolia* and *Antennaria rosea),* western anemone *(Anemone occidentalis),* alpine poppy *(Papaver kluanensis),* blue-bottle gentian *(Gentiana glauca),* Lyall's saxifrage *(Saxifraga lyallii),* moss campion *(Silene acaulis),* white dryas *(Dryas integrifolia),* and yellow paintbrush *(Castilleja lutescens).*

British Columbia Ranges

The Continental Divide was used by the Canadian government as the boundary between the southern parts of British Columbia and Alberta. So everything in this part of British Columbia is on the western slope of the Divide. As such, the environmental conditions and vegetation of this region have much in common with the adjacent west-slope regions of northern Idaho and Montana. The Main Range is the corridor of high mountains that straddle the Continental Divide. West of this the southern Rocky Mountain Trench separates the western Main Range from the Purcell Mountains, the Columbia Mountains, and the Monashee Mountains.

This is generally a wet region that receives Pacific moisture year-round, but the amount of precipitation varies greatly from one locality to the next. The Rocky Mountain Trench is a broad depression separating the western Main Range from ranges farther west. The plant life of the trench is quite diverse, owing to the variety of annual precipitation there. The driest parts of the trench are similar to the dry eastern slope of the Rockies in Montana. Some of the northernmost populations of prickly pear cacti *(Opuntia)* grow in this relatively warm, dry valley. Ponderosa pine parkland occurs from the Columbia Lake region to the international border. Farther north, the trench is heavily wooded with Pacific Northwest forest that includes Douglas-fir, spruce, western red cedar *(Thuja plicata)*, western hemlock *(Tsuga heterophylla)*, and a subspecies of balsam poplar called black cottonwood *(Populus balsamifera trichocarpa)*. This forest clothes the montane zone of the ranges to the west, notably the Columbia Mountains (the regional name for this forest is "Columbian Forest"). It is also sometimes called cedar-hemlock forest. These forests also contain other conifers, including white spruce, black spruce, lodgepole pine, and white pine, in addition to other deciduous trees, including Pacific willow *(Salix lasiandra)* and green alder *(Alnus viridis crispa)*. The understory of this type of forest includes the spine-covered shrub called devil's-club *(Oplopanax horridum)*, the medicinal plant, heal-all *(Prunella vulgaris)*, which might be useful if you run into a patch of devil's-club, as well as orange honeysuckle *(Lonicera dioica)*, wild sarsparilla *(Aralia nudicaulis)*, and thimbleberry *(Rubus parviflorus)*. This is lush vegetation, nurtured by lots of rain and a relatively mild climate year-round.

Montane forests in the Rockies of British Columbia receive more moisture than their eastern-slope counterparts; the increased moisture allows more luxuriant growth of understory vegetation and denser stands of trees. The forests of this zone include Douglas-fir and lodgepole pine, but also have the more moisture-loving western larch *(Larix occidentalis)*. Engelmann spruce and subalpine fir are the dominant trees of the subalpine forests in this region, and these species range down into western larch stands in the Kootenay region. Western larch does best in regions that have been disturbed by fire or logging; this tree grows faster than any other conifer in the region, but old stands are most often being colonized by other tree species, such as Douglas-fir or Engelmann spruce.

The northern edge of the Rockies intergrades with the boreal forest region. The boreal forest dominates the lowlands of northern Alberta and British Columbia, the southern and central Northwest Territories, southern and central Yukon Territory, and central Alaska. Black and white spruce are the principal conifers of this enormous forest region. The climate in many parts of the boreal forest is very continental, with bitterly cold winters and surprisingly warm, brief summers.

Northern Montana and Idaho Ranges

The northern Montana and Idaho ranges have some of the most spectacular scenery in the region. There are lakes both large and small, mountains with tremendous relief, waterfalls, glaciers, and glacially carved U-shaped valleys, all clothed in lush vegetation. The vegetation patterns adhere closely to the climate pattern of the region, which is far more moderate and moist on the western slope and far more dry and continental on the eastern slope. Western-slope climate in this region features milder temperatures year-round and a frost-free season of as much as eighty days. Mean annual precipitation ranges from 24 to 37 inches (610 to 940 mm). In contrast, the eastern-slope climate in this region features sharper differences between winter and summer temperatures and a frost-free season of only thirty days. Mean annual precipitation in this region is about 15 to 23 inches (380 to 580 mm) Because of these differences, the vegetation of the eastern slope is similar to other eastern-slope regions of Montana, and the western-slope vegetation has many species in common with the Pacific Northwest region.

Because of the sharp differences in composition between forests on the eastern and western slopes of this region and the complexities of forest stand composition on the western slope, what I present is, once again, a series of generalities that cover the whole region.

In our hypothetical tour of this part of the Rockies, we will begin on a hillside 3,300 feet (1,000 m) above sea level, a good average elevation for the lower forest boundary. Western-slope lowlands are dominated by sagebrush steppe, and eastern-slope lowlands have short-grass prairie with only a smattering of sagebrush. These treeless lowlands account for only about 10 percent of the region, however. The pines we are wandering through are Ponderosa pine *(Pinus ponderosa)*, a species we will see in almost every lower montane forest in the south. The grass that covers most of the ground is Idaho fescue *(Festuca idahoensis)*.

As we hike up the hill, you will notice that the Ponderosa pine gives way to Douglas-fir. If you look over to the rocky ridge, you will notice a little stand of whitebark pine. The higher elevations of these montane forests also have stands of Englemann spruce and subalpine fir, especially on cool, moist, north-facing slopes. On the western slope, where the climate is moister, I could show you stands of some of the Pacific Northwest trees, such as western larch, western hemlock, western red cedar, western white pine, and a dwarf form of Pacific yew *(Taxus brevifolia)*. These trees need at least 32 inches (810 mm) of precipitation per year.

There is nothing too remarkable about the subalpine forests of this region—just your typical mix of Englemann spruce and subalpine fir in most locations, growing up to 8,500 feet (2,600 m). There is also some subalpine larch that grows in moister patches. Larches shed their needles in winter, so they can often grow at higher elevations than other kinds of conifers, because they are better able to withstand the scouring effect of wind-blown ice crystals. We tend to think that all conifers are evergreens, but larches are an exception.

I've brought you up to Logon Pass today, in Glacier National Park. This is one of the best places to get a good look at the alpine tundra zone of northwestern Montana. We have been driving up the Going-to-the-Sun Highway, surely one of the most spectacular roads in the Rockies. The alpine wildflowers form a brilliant layer of color on top of the rich tundra vegetation of this well-watered region. Take your time looking at these miniature marvels, and don't forget to get down on all fours now and then for a closer look at the carpet of tiny petals that sits only an inch or so off the ground.

Montane Pacific Northwest forest, Glacier National Park, Montana

Alpine meadow, Logan Pass, Glacier National Park, Montana

West-Central Montana and East-Central Idaho Ranges

The west-central Montana and east-central Idaho ranges region includes the northern half of the Bitterroot Mountains, the Mission Range, the Swan Range, the Front Range (also called the Sawtooth Range), the Big Belt Mountains, a few smaller ranges in Montana, and the Clearwater Mountains of Idaho. The Bitterroot Range forms a significant barrier to Pacific moisture. The ranges to the east are in its rain shadow, so they are considerably drier. Pacific Northwestern plant species are therefore relatively rare in this region. They are found only in moist canyon floors or seepage areas. Grand fir *(Abies grandis)* grows here, but it is much less common than it is in the wetter regions of northwestern Montana and northeastern Idaho, where Pacific moisture is much more abundant. About 75 percent of the land in this region of Idaho and Montana is forested. Broad, low-elevation valleys have grassland dominated by bluebunch wheatgrass *(Agropyron spicatum)*, Idaho fescue, and rough fescue *(Festuca scabrella)*. The average elevation of lower treeline in this region is between 3,200 and 5,500 feet (975 and 1,675 m), rising higher on warm, dry slopes. The average elevation of upper treeline is 8,800 feet (2,690 m), and few peaks in the region are much higher than this, so there is little alpine tundra in west-central Montana.

The forests of this region are botanically similar to those of southwestern Montana and southern Idaho. The dominant montane forests are mainly Douglas-fir, and the dominant subalpine forest is mainly subalpine fir. The understory species composition is quite similar to that found farther south, except that the milder, moister climate in this region has allowed a few Pacific Northwest elements to flourish.

Subalpine larch grows with subalpine fir in the upper subalpine region on the western slopes of this region. Near the upper tree limit, almost pure stands of subalpine larch grow above the elevational limit of other tree species.

Southwestern Montana and Southern Idaho Ranges

The mountains in the southwestern Montana and southern Idaho ranges include the Pioneer Mountains, the Anaconda Range, the Sapphire Mountains, the Tobacco Root ranges of Montana; the Bitterroot Range that straddles the border between Montana and Idaho; and the Salmon River Mountains, the Sawtooth Range, the Boise Mountains, the Boulder Mountains, the Lost River Range, and the Lemhi Mountains in Idaho. Some of the mountains are separated by broad valleys covered with grassland; others are separated only by steep, narrow canyons. This is a cold, dry forest region with high valley floor elevations. The climate is continental. Annual precipitation ranges from about 10 to 50 inches (250 to 1270 mm). The growing season is evidently too short to allow the establishment of Ponderosa pine in this part of Montana, but it grows in most of the ranges in Idaho because of the somewhat milder, wetter climate there.

Despite its mountainous terrain, only about a quarter of this region is forested. Grasslands cover the lowlands at elevations up to about 5,000 feet (1,600 m). In the driest valleys and slopes, herbaceous vegetation ranges up to about 6,000 feet (2,010 m). The lowland vegetation is dominated by semiarid steppe with big sagebrush and grasslands.

In Montana, the lower tree limit is formed by stands of limber pine, growing in open parkland. In Idaho, it is formed by Ponderosa pine. As you will see as we hike up the hill, the trees only grow to a height of about 20 feet (6 m), and the appearance of this forest is reminiscent of pinyon–juniper woodlands in the southern Rockies. That is no coincidence. They grow in the driest regional habitats supporting the growth of conifers. The shrub with the yellow berries and waxy leaves that you see everywhere is called Oregon grape *(Mahonia repens)*. It has no botanical relationship with real grapes, and the berries are not palatable to people, but I find them attractive anyway.

Douglas-fir dominates the montane forests of this region. Depending on moisture conditions, these trees range up to 7,500 feet (2,300 m). Douglas-fir forests and subalpine fir forests are the two most important forest communities in Idaho and Montana, covering more mountainsides than any other group. Rocky slopes often have an understory of common juniper, but cold, moist slopes have an understory dominated by huckleberry *(Vaccinium)*. Unlike Oregon grape, huckleberry shrubs produce a very tasty, highly palatable fruit. This is a regional delicacy, harvested by both amateur and professional berry pickers for use in syrups, jams, jellies, and baked goods.

A narrow zone of lodgepole pine dominates the slopes just above the Douglas-fir forests in this region. This zone rarely spans more than 300 feet (100 m) of elevation, but lodgepole pine intergrades with forests above and below this range, especially following disturbances such as fires.

In the southwestern part of the state, subalpine fir-dominated forests range from

7,000 feet (2,130 m) on cool, moist slopes to 8,350 feet (2,550 m) on drier slopes. Above relatively pure stands of subalpine fir, whitebark pine joins the composition to form the upper treeline forest. Whitebark pine suffered serious damage from bark beetle *(Scolytidae)* outbreaks in this region during the past eighty years, and larger individual trees have almost been eliminated from most regions. Treeline in this region ranges from about 8,700 to 9,500 feet (2,650 to 2,900 m).

Plant Communities of the Central Rockies

The central Rockies comprise a series of relatively isolated ranges in northeastern Utah, central and northern Wyoming, southwestern Montana, and southeastern Idaho. The southern and western ranges of the central Rockies are bordered by sagebrush steppe, and the northeastern ranges are bordered by shrub steppe with wheatgrass and needle-and-thread grass *(Stipa comata)*. These dry lowland regions receive less than 10 inches (250 mm) of annual precipitation. Although many parts of the central Rockies are relatively dry, the climate of mountain ranges is also quite cold. Mean July temperatures are generally less than 67°F (19°C). The mountains of the central Rockies serve as refuges for moisture-loving plants. Approaching these mountain ranges from almost any direction necessitates hours of driving through unbroken expanses of arid steppe, often dominated by sagebrush. However, the reward at the end of the journey is particularly gratifying to the senses. The deep greens of the forests and the profusion of running and standing waters that await the visitor in the mountains make the memories of the trip through the high plains fade away like a mirage.

Beartooth Mountains

The Beartooth Mountains are essentially a northern part of the Absaroka Range that straddle the Wyoming–Montana border northeast of Yellowstone National Park. This region is unique in the central Rockies because most of it is so high. The Beartooth Plateau that forms the eastern half of the region is mostly above 10,000 feet (3,050 m), placing it well above regional treeline. There are about a dozen other high plateaus in these mountains, separated from each other by deep canyons carved into U-shapes by glacial ice. There are many peaks above 12,000 feet (3,650 m) in the Beartooth Mountains; glacial and periglacial landscapes are common. The Beartooth Mountains received a protected status in 1978 when about 900,000 acres (328,000 ha) became the Beartooth-Absaroka Wilderness. This section will focus on alpine tundra, because that is what makes the Beartooth region so special.

The climate of the alpine zone in the Beartooth Mountains is as cold as some arctic regions. Afternoon temperatures in the summer months average only about 48°F (9°C), and many nighttime temperatures fall below freezing. The wettest, least windy sites are dominated by snow beds, which persist into late July and August, leaving behind soils saturated with meltwater for a few brief weeks before snowfall resumes (September or October). These wet patches of the tundra have icegrass *(Phippsia algida)* and Iceland plant *(Koenigia islandica)*, two arctic and alpine species adapted to wet, gravelly substrates.

Bogs dominated by sedges *(Carex)* develop on wet soils that are unusually high in organic matter. These sites are found in shallow basins that are downhill from snow beds. These depressions are fed by meltwater throughout the summer, so they never dry out. Cotton grass also grows in these bogs, as does marsh marigold *(Caltha leptosepala)*, American bistort *(Polygonum bistortoides)*, and saxifrage.

Meadows dominated by tufted hairgrass *(Deschampsia caespitosa)* grow in shallow basins and higher up where there is good winter snow cover. The snow cover protects plants in these localities from the vicissitudes of winter climate. These meadows are often disturbed by the burrowing of pocket gophers. These industrious little rodents can turn an alpine meadow into gravel mulch with a few scattered plants. Another main source of disturbance in this type of plant community comes from frost action in the soil. Frost-heaving uproots plants, leaving bare mineral soil and piles of rocks. Frost action can also create hummocks. These are areas that are raised above the surrounding landscape, creating a series of alternating mounds and troughs.

Alpine avens *(Acomastylis rossii)* turf forms on ridges that are mostly blown free of snow in winter. The cushion plants on the dry ridges include tufted phlox *(Phlox caespitosa)*, moss campion, alpine clovers *(Trifolium)*, and spikemoss *(Selaginella densa)*. False elk sedge *(Kobresia myosuroides)*, the dominant turf-forming species on the alpine tundra of the southern Rockies, is only found on windward slopes of the Beartooth Plateau.

As might be expected in such a large, isolated oasis for tundra species, several species of plants grow that are not found elsewhere in the region. There are also several species thought to be endemic to this region—that is, they exist nowhere else. These include Dodge willow *(Salix rotundifolia dodgeana)*, Limestone columbine *(Aquilegia jonesii)*, palish larkspur *(Delphinium glaucescens)*, Hayden's clover *(Trifolium haydenii)*, pretty dwarf lousewort *(Pedicularis pulchella)*, slender fleabane *(Erigeron gracilis)*, and Rydberg's daisy *(Erigeron rydbergii)*.

Vegetation of Northwestern Wyoming

The northwest corner of Wyoming and adjacent regions of Idaho and Montana encompass a variety of mountains and highlands famous for their scenic beauty and variety. At the southern end of this region is Grand Teton National Park, in which the Snake River Valley runs between the Gros Ventre Range on the east and the Teton Range on the west. To the north, Yellowstone National Park is a highland region nearly ringed by mountains. On the east side of the park, the Absaroka Mountains form a chain that arcs north into Montana. The northwestern region of the park has the Gallatin and Madison ranges, and the northeast corner is bounded by the Beartooth highlands.

Yellowstone and Grand Teton National Parks

The vegetation of Yellowstone and Grand Teton National Park is essentially the same, except for differences caused by varying types of soils. Whereas the Jackson Hole region and adjacent mountains are dominated by either granitic or calcareous substrates, the Yellowstone region is dominated by volcanic soils. These soils are unsuitable for any of the regional tree species except lodgepole pine. Most of Yellowstone is, not surprisingly, covered by lodgepole pine forest.

In addition to the central plateau lodgepole pine forest, some other plant communities typify the Yellowstone region but are rare or absent from the Teton region.

These include both montane and subalpine forests. The Yellowstone region has a high plateau nearly surrounded by highlands and mountain peaks. The southern part of the park is higher than the northern part and receives considerably more precipitation.

The Engelmann spruce forests of northeastern Yellowstone are unlike any in the Teton region. These occur mostly in the subalpine zone, from 6,200 and 8,200 feet (1,900 and 2,500 m). Engelmann spruce is the only stable tree in the forest, whereas lodgepole pine and Douglas-fir come in after a disturbance. The understory is often clothed by a carpet of twinflower *(Linnaea borealis)*.

A subalpine fir community dominates many moist habitats between 6,800 and 9,100 feet (2,070 and 2,775 m) in Yellowstone. High-elevation forests of Yellowstone that approach treeline are dominated by whitebark pine.

For many years, ecologists working in Yellowstone National Park have observed that fire suppression has altered the landscape. It is difficult to fully assess the effects of fire suppression on the vegetation patterns of a landscape. For one thing, we do not have the luxury of comparing large, undisturbed regions with regions where fire suppression has taken place, because fire suppression has been in effect during the past century or more in just about every region of the Rockies. Fires were suppressed in the Yellowstone region from 1886 to 1975. Before then, fossil evidence indicates that fires burned patches of Yellowstone forests on an average of every twenty-five years. On the whole, human interference with this natural fire cycle has led to decreased diversity of regional biological communities. This may seem counterintuitive, because we often view fire as an agent of destruction, but fire is nature's tool for thinning forests. Without it, forests become substantially more dense, and the dense stands block out enough light to inhibit the growth of many understory species of shrubs and herbs. The abundance of aspen, the noted successional species in regional fire disturbances, was greatly reduced in the twentieth century. Based on comparisons of old and recent photographs, sagebrush parkland has been expanding at the expense of more diverse communities with herbs and other shrubs. Stream-side willow and alder thickets have been reduced in size and density during this fire-free period, as well. Fire appears to be essential in maintaining the greatest number of species in Yellowstone mainly because it gives successional species (those that come in after disturbance) an opportunity to flourish.

The role of fire in initiating and perpetuating lodgepole pine forests has been discussed by forest ecologists throughout the past fifty years. Initially it was considered only a successional species that came in after fire or other disturbance, but additional studies revealed long-lived, stable lodgepole pine communities, such as the forests growing on the central plateau of Yellowstone. The question remained, however, "Does lodgepole depend on fire to maintain itself?" A study in Yellowstone indicates that this is not the case. Even though fire has occurred throughout the park, most lodgepole pine stands are not maintained by fire. The fire ecologists got a chance to test their theories in a big way in 1988, when about half of the park was burned by massive forest fires. By that time, the National Park Service had already decided not to suppress natural fires in the park.

Jackson Hole is a deep valley that forms the lowland region of Grand Teton National Park. Fur trappers in the early nineteenth century named Rocky Mountain Valleys surrounded by mountains "holes." This region features long, cold winters

Lodgepole pine forest, Yellowstone National Park, Wyoming

NANETTE ELIAS

Forest fire in Grand Teton National Park, Wyoming. This kind of lightning-strike fire led to the major fires that burned through half of Yellowstone National Park in 1988.

with deep snow accumulations. The elevations of the park are all above 6,000 feet (1,800 m). There is a short growing season and low mean annual temperatures. Jackson Hole is bounded by the Gros Ventre Range on the east, the Absaroka Range on the north and northeast, and the Teton Range to the west.

The rolling hills above the meadows in Jackson Hole are dominated by a big sagebrush community. Lowland meadows in this region are dominated by two plant communities. One of these is dominated by sedges and the other is a sedge-grass mixture. Riparian (streamside) forests in this region are dominated by narrow-leaf cottonwood *(Populus angustifolia)* and balsam poplar. You may spend a good deal of time in one or another of these kinds of vegetation if you camp in the park, because several campgrounds were built there. The riparian region is also favored by moose *(Alces alces)* who have a weakness for willow leaves. Scattered clumps of blue spruce *(Picea pungens)* are also found. This is one of the lowest elevation spruces of the Rockies, found down to the foothills in many regions, where water is abundant.

The forest zones in Grand Teton National Park begin with limber pine parklands from 4,900 to 7,875 feet (1,500 to 2,400 m). Above the limber pine zone is a narrow band of Douglas-fir forest. This community ranges from about 6,200 to 6,600 feet (1,900 to 2,000 m). Lodgepole pine is the dominant forest type between the Douglas-fir forests below and the subalpine spruce-fir forests above. Its main elevational range in the Teton region is from 6,560 to 7,875 feet (2,000 to 2,400 m). In northern and western Wyoming, from the Wyoming and Salt River Ranges to the Wind River, Gros Ventre, and Teton Ranges, this species demonstrates a broad ecological amplitude. It spans the environmental range between the cold part of the Douglas-fir habitats and all but the wettest parts of the subalpine fir-Engelmann spruce habitats. Thus this species is found in every forest zone except the most dry, where only limber pine can grow.

Subalpine forests of Engelmann spruce, subalpine fir, and whitebark pine grow at elevations from 7,875 feet (2,400 m) to upper treeline at 9,350 feet (2,850 m). The upper treeline stands are dominated by whitebark pine. Engelmann spruce is the most successful, occupying a broader range of habitats than subalpine fir.

The slopes of the Teton Range are very steep, especially the eastern-facing slopes. Avalanches are rather common on these slopes, and the paths where avalanches repeatedly occur have some interesting plant communities. Tracks where avalanches occur often tend to have great densities of shrubs and conifers. The size of the conifer trees is reduced in these tracks, because being a tall tree is a liability there. Slow growth appears to be a successful adaptation to life in a dangerous locality, so much so that these avalanche plant communities seem to be quite stable through time, in spite of the massive disturbance wrought by repeated avalanches.

Wind River Range

The Wind River Mountains are the highest range in Wyoming, with an extensive alpine tundra zone above 10,000 feet (3,000 m). Gannett Peak, at 13,804 feet (4,207 m), is the highest point in the state. The range trends northwest–southeast, and precipitation patterns differ on the eastern and western slopes. On the western slopes the precipitation is relatively even throughout the year, whereas on the eastern slope, precipitation peaks in May, summer precipitation is somewhat higher than on the eastern slope, and winter precipitation is very low. Unfortunately,

Subalpine forest, Grand Teton National Park, Wyoming

few virgin stands of trees are left in the Wind River Range, because the region was extensively logged during the nineteenth century.

Douglas-fir stands form the lower treeline. On steep, north-facing slopes this species ranges down to 7,200 feet (2,200 m) elevation, where it meets sagebrush steppe and grasslands. The ecotone (boundary) between these two communities is often occupied by stands of aspen or limber pine. The upper elevational limit of Douglas-fir is about 8,500 feet (2,600 m). Lodgepole pine dominates most regions at elevations between about 7,900 and 9,500 feet (2,400 and 2,900 m). It is best developed on the southern third of the range.

Elevations between about 9,500 and 10,300 feet (2,900 and 3,150 m) are dominated by subalpine forest of Engelmann spruce and subalpine fir, although subalpine fir also ranges downslope to elevations as low as 7,900 feet (2,400 m) on north-facing slopes. Whitebark and limber pines also grow in the subalpine zone, on exposed ridges and other windswept localities.

The alpine tundra zone of the Wind River Range extends up from the edge of the subalpine forest where dwarf forms of spruce and fir grow. These short, twisted-trunk trees are called krummholz. They do their best to hug the ground, keeping out of the icy blasts of winter winds.

Big Horn Mountains

The Big Horn Mountains of north-central Wyoming are the other mountain region that is relatively isolated from other highlands in the central Rockies. It is in many ways typical of the central Rockies, but it also has some unique features because of its isolation and unique precipitation pattern. This range receives relatively little precipitation (20 to 30 in. or 510 to 760 mm per year), but unlike most other ranges east of the Continental Divide, the Big Horns receive more moisture on their eastern than on their western slopes. This precipitation comes from the Gulf of Mexico. The Big Horn Mountains rise from a base of about 3,000 to 4,000 feet (900 to 1,200 m) to a height of more than 13,000 feet (4,000 m) at Cloud Peak. The basin to the west of the Big Horns is one of the driest regions of the Rockies, receiving an average of only 6 inches (150 mm) of precipitation per year.

About 60 percent of the Big Horn region is forested. The lower limit of trees occurs at about 4,900 feet (1,500 m) and the upper limit occurs at about 10,000 feet (3,050 m). The range is surrounded by lowlands covered by grasslands with sagebrush, and big sagebrush *(Artemisia tridentata)* grows right up to the mountaintops in some regions. The lowest forest stands on the mountains are Ponderosa pine. These forests range from 5,000 to 6,000 feet (1520 to 1830 m). Above this, Ponderosa pine gives way to Douglas-fir. The Douglas-fir zone has been heavily lumbered during the past century, and very few undisturbed stands remain.

The most important conifer in the Big Horns is lodgepole pine. These forests are best developed in the center of the range where granite bedrock is exposed over large areas. The lodgepole pine forest is the first closed forest encountered in the Big Horns, as the Douglas-fir and Ponderosa pine forests below are mostly parkland. The rain shadow southeast of the central peaks of the range allows large areas of grassland in the lodgepole pine zone. Though stands of lodgepole pine range up to treeline on granite substrates, the main zone falls between 7,000 and 9,500 feet (2,130 and 2,900 m).

The subalpine forests of the Big Horns are the typical mix of Engelmann spruce and subalpine fir. These forests dominate the land from 8,500 to 10,000 feet (2,600 to 3,050 m), especially on north-facing slopes. Lodgepole pine and Douglas-fir also encroach on the subalpine zone. Mosses and lichens cover much more open ground than herbs and shrubs in this forest.

Because the Big Horns are among the drier ranges of the Rockies, herbs and shrubs play a much larger role than in most other regions. On the western slope of the range, Utah juniper *(Juniperus osteosperma)* and mountain mahogany *(Cercocarpus montanus)* communities are important, and big sagebrush is found throughout the range, exceeding 40 percent cover near upper treeline. The Big Horns are a sort of botanical crossroads of the Rockies. This is the southernmost range of the Rockies that has whitebark pine and the northernmost range that has Utah juniper.

Uinta Mountains

The Uinta Mountains are unique in the Rocky Mountain region because they are the only east–west trending range. A broad, high upland region of alpine tundra caps the range. Peaks in the Uintas reach elevations of 13,000 feet (3,960 m). The north slope of the range presents an extremely steep face; in some places the north slope is almost vertical. The southern slope is much more gradual.

The vegetation of the Uinta Mountains shows strong relationships with the regional bedrock geology. This region borders the Great Basin on the west, and much of the Uintas receive less than 20 inches (510 mm) of precipitation per year. Montane forests that grow on limestone substrates are dominated by Douglas-fir, whereas montane regions with quartzite substrates support forests of lodgepole pine. In the subalpine zone, subalpine fir is nearly absent from quartzite substrates, except in very wet soils. It dominates subalpine regions underlain by shales. Engelmann spruce does well on both limestone and quartzite substrates. The alpine vegetation pattern is also strongly influenced by substrate type. Alpine avens is nearly ubiquitous on quartzite substrates in the alpine but practically absent from limestone regions. Alpine willows thrive on the limestone substrates of the alpine in the Uintas.

Ponderosa pine forests are found in the montane zone of the eastern Uintas. Lodgepole pine forests range from about 8,000 to 9,500 feet (2,440 to 2,900 m). These stands are associated with regions of high fire frequency.

Subalpine forests range from about 9,000 to 11,000 feet (2,750 to 3,350 m) in the Uintas. Engelmann spruce tends to dominate stands above about 10,400 feet (3,170 m). Unlike other ranges of the central Rockies, five-needled pines such as limber pine do not grow near the upper treeline in the Uintas. Grouse whortleberry *(Vaccinium scoparium)* is the dominant understory plant throughout the subalpine zone. Subalpine meadows are also a common feature of this range. Tufted hairgrass *(Deschampsia caespitosa)* and timber oatgrass *(Danthonia intermedia)* are common herbs in these meadows.

The alpine tundra zone of the Uintas forms an unbroken region of about 300 square miles (780 km^2). This is a rather unique topographic setting for the alpine zone in the Rockies. The lower limit of tundra falls at about 11,000 feet (3,350 m), where krummholz forms of spruce and fir give way to herbs. Wind-swept ridges and summits in this region support communities of rock sedge *(Carex rupestris)* and a variety of cushion plants, including moss campion, Rocky Mountain nailwort *(Paronychia pulvinata)*, and alpine clover. Areas less exposed to wind are covered by alpine avens and false elk sedge.

Plant Communities of the Southern Rockies

We will begin our tour of the southern Rockies at the top of the mountain, in the alpine tundra. After all, it is easier to hike downhill than up. Above about 11,500 feet (3,500 m), subalpine forest gives way to alpine tundra in the southern part of the southern Rockies. The upper limit of spruce and fir individuals is often higher, around 12,000 feet (3,650 m). In the central and northern parts of the southern Rockies, the boundary between subalpine forest and alpine tundra begins at about 11,000 feet (3,350 m) and ends at about 11,500 feet (3,500 m). In between these elevations there is often a broad boundary zone, or ecotone, between the two communities. This ecotone varies from slope to slope and from mountain to mountain, depending on topographic variations, substrates, steepness of slope, and aspect.

Treeline is a tough place for trees to survive. The high winds and chilling temperatures of these high elevations are devastating to upright trees. The effects of severe climate can be seen a little lower downslope, where the tops of upright trees have been stripped of needles and bark, leaving bare wood bleaching white in the sun. Stands of such trees growing near treeline look as if they are waving white flags of surrender to the alpine climate. It is not high winds alone that strip the needles and bark from trees too close to treeline but the effect of ice crystals carried by the winds. Hundreds of years ago, a prolonged period of warm climate allowed trees to get established well above the modern tree limit. When the climate cooled again, they died out. All that is left of these trees is dead, decaying trunks, sitting like white specters in the alpine tundra. This is an ecological chess game between vegetation and climate in which the two contestants take centuries to complete their moves. The alpine climate, however, is an unforgiving opponent, and conifer seedlings sprouting above the established treeline are killed most of the time. Patches of prostrate krummholz trees form tree "islands."

Clumps of prostrate conifers that form tree islands do not remain in place in their tundra "sea." They creep across the landscape, a few centimeters per year, always in the leeward direction of the prevailing winds. This is accomplished by horizontal branches taking root on the leeward edge of the plant. The upwind edges are gradually killed by ice crystal-blasting and extreme drying.

The alpine tundra has so many special qualities that are difficult to capture on the printed page. Even from a distance you cannot appreciate the tundra. What the wildflowers lack in stature they more than make up for in dazzling color. The tiny petals of moss campion and alpine forget-me-not are brilliant pink and blue; the alpine phlox *(Phlox pulvinata)* flowers are a brilliant white against the dark green background of tiny leaves. After you've spent some time there, you will get the feeling of being let in on a wonderful secret: The tundra is not just a bleak, treeless landscape half-buried by ice and snow; there is great beauty, if you look closely enough. This is my favorite patch of the Rockies, where the inhabitants succeed against staggering odds and manage to look pretty doing it.

The tundra growing season in northern New Mexico is the longest of any region of the Rockies, averaging ninety days in June, July, and August. Farther north in the southern Rockies, the growing season averages about seventy days, but some localities are covered with semipermanent snow beds, making the growing season at these sites effectively zero. Although mean annual precipitation may exceed 35 inches (900 mm), the almost constant alpine winds redistribute the snow so that deep snow beds accumulate in hollows and windswept areas are quite dry. Alpine air is often dry, anyway. This, combined with high winds and strong insolation, makes for severe *evapotranspiration* (the sum of water loss from the landscape through evaporation and plant water loss).

The alpine tundra habitats include herbaceous meadows, rocky talus slopes, snow beds and adjacent moist areas, and tundra ponds. Because of the great topographic diversity of mountaintops, these habitats are rarely formed in large unbroken swaths. Rather, they exist as small patches.

Fellfields are, for the most part, bare, rocky slopes above treeline. There is no soil development on most fellfield slopes, so plants have to colonize in the nooks and crannies where soil has collected. The meager soils are easily eroded from their rocky substrates, and they hold little water. These regions are severely cold and dry, even by alpine tundra standards. Fellfield communities are made up mostly of cushion plants. These are mostly perennials that grow low to the ground in the shape of a flattened pin cushion, an aerodynamic shape that offers the least wind resistance. Common species in fellfield communities include alpine avens, alpine forget-me-not, alpine clovers, moss campion, and alpine sandwort *(Minuartia obtusiloba)*.

On many gentle alpine slopes, the vegetation is dominated by the mat-forming species, mountain dryas *(Dryas octopella, Dryas integrifolia)*. The plants obstruct the creeping soil, causing the formation of small terraces. This is one alpine plant that takes advantage of an otherwise disturbed soil to make itself a place to grow.

False elk sedge meadows clothe gently rolling hills and level surfaces of the alpine in the southern Rockies. False elk sedge is not a true sedge (genus *Carex*), even though it is a member of the sedge family (Cyperaceae). It grows as a thick turf with intertwining roots, often excluding other species. Sometimes interlopers get a foothold, such as alpine avens. The avens' leaves turn bright red in autumn, contrasting with the coppery-golden hues of the false elk sedge at that time of year. The best developed true grass meadows of the alpine tundra are stands of tufted hairgrass. Like false elk sedge, this bunchgrass grows in very tight clumps that seemingly exclude other species. Unbroken fields of tufted hairgrass cover large regions of the tundra.

At the wet end of the scale are alpine snowbank communities. These lie under and next to snow banks that persist for most if not all of the year. Melting snowbanks frequently support dwarf willows, less than 2 inches (5 cm) tall, as well as herbaceous vegetation such as Drummond rush *(Juncus drummondii)*, tufted hairgrass, and Coulter daisy *(Erigeron coulteri)*. Less wet patches of soil support taller willow scrub species that grow up to 3 feet (1 m). Stream-side communities that form along meltwater channels include groundsels *(Senecio)*, kings crown *(Sedum integrifolium)*, rose crown *(Sedum rhoanthum)*, and Parry primrose *(Primula parryi)*.

Many tundra plants have remarkable adaptations for survival in a cold, harsh climate. Many species have leaves and stems covered with a layer of pale hairs, which insulate the plants from the drying and chilling effects of winds by slowing the air down just before it hits the plant tissues. A good example of this "hairy" adaptation is old man-of-the mountain. The insulating hair layer also helps the plants hold the warmth coming from sunlight, so that they can begin the chemical reactions of photosynthesis in lower air temperatures. The dead air layer created by the hairy fibers also helps prevent moisture loss, another critical element in the alpine zone, where the air can be extremely dry.

Many alpine plants are adapted to perform photosynthesis at temperatures near the freezing point. Although plants of lower elevations are biochemically dormant in near-freezing temperatures, tundra plants are busy converting solar energy into stored chemical energy in the form of carbohydrates. Most alpine plants are much bigger below ground than above. They use massive roots and rhizomes to store energy while keeping their above-ground parts as short as possible to avoid the drying, chilling effects of wind.

Solar energy in the alpine is a vital plant resource, both for warming of tissues and for photosynthesis, so some plants have flowers shaped like parabolic disks to catch as much sunlight as possible. But no matter how efficiently the alpine plants use solar energy, one of their toughest obstacles to overcome is the shortness of the growing season. Winter cold and snow dominate the alpine zone from about October through May, and frost occurs even in the summer months. It is nearly impossible for plants to fulfill their life cycle in the space of a few short weeks in summer. That is why nearly all alpine plants are perennials. Perennials live more than one year, and some alpine plants live more than a decade. This longevity allows plants to slowly accumulate nutrients in their roots and rhizomes, summer after summer after summer, until they finally have enough extra energy to develop a crop of viable seeds. And alpine perennials have another trick up their sleeve. Many produce preformed shoots that lie dormant in winter but burst into action as soon as the snow melts, jump-starting their summer activity.

Subalpine Zone

Subalpine forests in the southern parts of the southern Rockies generally range from 8,500 to 12,000 feet (2,600 m to 3,650 m) on north-facing slopes and from 10,000 to 12,000 feet (3,050 to 3,650 m) elsewhere. The two dominant tree species are—you guessed it—Engelmann spruce and subalpine fir. The climate of the subalpine zone in the southern Rockies is characterized by long, cold winters with lots of snow, followed by short, cool summers. The average January temperature in this region ranges from 20°F (-°C) in northern New Mexico to 15°F (-9°C) in southern Wyoming. Average July temperatures range from 60°F (16°C) in northern New Mexico to 50°F (10°C) in southern Wyoming. The frost-free period ranges from sixty days in the south to thirty days in the north, but frost can occur during any month of the year. Average annual precipitation varies from 24 to 35 inches (610 to 890 mm). Most of this comes as snow, which averages almost 204 inches, or 17 feet (5,000 mm) each winter. Spruce-fir forests are often buried in deep snows for most of the year. They are adapted to greater moisture than the conifers of the montane forests, and they cannot tolerate dry soils for long.

The slim, conical shape of these trees prevents their branches from accumulating significant amounts of limb-breaking snow. The trunks of trees in the lower subalpine elevations are fairly uniform, thinning from bottom to top, but the trunks of trees near treeline are substantially thickened at the base. Such trunks are resistant to being broken in high winds that come roaring off nearby mountain tops. The subalpine environment is cold and harsh, and trees grow extremely slowly. There is a sort of biological pace associated with different ecosystems. In general, the warmer the temperatures, the quicker the pace. For instance, on the prairies east of the Rockies, individual plants (mostly herbs) regenerate year by year, and recover from fires in a year or less. In very cold climates, such as the subalpine forest, the dominant species are extremely long lived, and recovery from disturbance takes centuries.

When it comes to slow growth, the champion tree and great granddaddy of the forest is the bristlecone pine *(Pinus aristata)*. On rocky subalpine outcrops in the southern Rockies, bristlecone pine trees can live several thousand years. Bristlecone pines have been the chief species used to develop long-time series in tree-ring research, because their rings may span the lives of many generations of other tree species. In the Rockies, bristlecone pine ranges from New Mexico to the southern parts of the Colorado Front Range.

The subalpine forests of the southern Rockies are some of the least disturbed of the region. They cover rough, steep terrain that has not allowed easy access, and their relatively low timber values have discouraged large-scale logging. The moist soils and late-lying snow tend to keep forest fires to a minimum. So, hiking through these stands is as close as we can get in this region to a walk through a primeval forest. The trees tower above your head, often reaching heights of 100 feet (30 m). In your pathway lie many dead snags, some with little bright green seedlings pushing their way up along side. Hence the forest cycle from birth to growth, maturity, and death are all plainly seen in one view. Unlike the montane forests, which are mostly open, sunny, and airy, the subalpine spruce-fir forests often have closed canopies, making them cool, shady, and quiet. These forests are a treat to all the senses.

When there is disturbance in the subalpine forest, aspen often comes in. Aspen is the only deciduous tree to grow at these elevations. Avalanches that start in the alpine zone most often end up crashing down into the subalpine, ripping up trees as they go. Violent wind storms hit patches of subalpine forest on occasion, toppling trees in their path. Rock slides and land slides are also not uncommon at these elevations, as water-saturated soils lose their grip on steep slopes.

Aspen excels at colonizing open ground. That is why it does so well in disturbed sites. Once it has become established, however, its ability to persist in a given region depends on a variety of factors. Where soils are moist and deep, and where there is a thick understory of shrubs and herbs, aspen can remain for many centuries. Where soils are relatively poor and dry, and the shrub and herb cover is poor, conifers tend to get established, eventually eliminating aspen from the site. A typical series of events in this succession would be as follows. A stand of subalpine fir and Engelmann spruce trees are burned in a fire caused by a lightening strike. Aspen seeds that were carried on the wind germinate in the charred ground that is left behind. The aspens grow tall and send out lateral shoots that sprout and form more trees. In this way one genetic individual can create a small stand of trees all by itself. The aspen branches shade the soil as the trees mature, allowing conifer seedlings from nearby stands to get established. After about 150 years since the initial colonization, spruce and fir seedlings have grown above the height of the aspen stand and begin to shade out the aspen leaves. The aspens start to die because of this, and spruce and fir trees take over, returning the forest to its predisturbance state.

The understory of spruce-fir that stands throughout the southern Rockies is a mixture of herbs and shrubs. The subalpine forests tend to grow tall and dense, shutting out most of the sunlight from the forest floor. As a consequence, the understory vegetation is limited to species that tolerate shade and are adapted to considerable moisture. The common subalpine groundcover of the southern Rockies is a mixture of shrubs and herbs, including huckleberry, bearberry *(Arctostaphylos uva-ursi)*, shrubby cinquefoil *(Potentilla fruticosa)*, twinflower, buffaloberry *(Shepherdia canadensis)*, forest fleabane *(Erigeron eximius)*, Jacob's ladder *(Polemonium pulcherrimum)*, strawberry *(Fragaria ovalis)*, fireweed *(Epilobium angustifolium)*, wood nymph *(Moneses uniflora)*, and monkshood *(Aconitum columbianum)*. This understory provides vital habitat for many mountain mammals. The variety and abundance of edible berries that grow in the subalpine offer tasty, calorie-rich foods that allow many animals to fatten up before the onset of winter.

Upper Montane Zone

We will continue our downward trek, leaving the subalpine, descending into the upper montane forests, which range from about 8,000 to 9,000 feet (2,450 to 2,550 m). The New Mexico and southern Colorado forests typically have Douglas-fir, white fir *(Abies concolor)*, several species of pine, blue spruce, and aspen. Annual

Montane forest, Front Range, Colorado

precipitation in these forests averages from 23 to 28 inches (600 to 700 mm) per year. In the central and northern parts of the southern Rockies, Douglas-fir and, to a lesser extent, Ponderosa pine dominate this zone. Botanically, the upper montane has much in common with lower montane forests, but the role of Douglas-fir makes it a distinct ecological community. The main difference is that Douglas-fir dominates even the south-facing slopes of the upper montane zone. The temperatures are just cool enough and precipitation sufficiently greater to give Douglas-fir dominance over Ponderosa pine. The understory of these forests includes bearberry (also known by its Native name, kinnikinnik), Rocky Mountain juniper, common juniper, wild raspberry *(Rubus idaeus)*, penstemon *(Penstemon)*, yarrow *(Achillea lanulosa)*, and cinquefoil.

The montane zones of the southern Rockies contain more than forested slopes and ravines. Relatively flat, broad regions ranging in size from small meadows to parks many miles across provide a variety of habitats for herbaceous vegetation. In these regions, the environment is not suitable for trees, but perennial herbs do quite well. These communities fall into a few ecological categories, including mountain grasslands and dry and wet meadows. One of the more interesting topographic features of the southern Rockies are the parks. These high-elevation, flat basins are semiarid grasslands that receive too little precipitation to allow the growth of trees, mostly because they lie in rain shadows from surrounding high mountains. Colorado has North Park, Middle Park, and South Park, as well as regions not called parks that are nonetheless high-altitude grasslands, including the San Luis and Wet Mountain Valleys. High-elevation grasslands also are found in southern Wyoming (the Shirley and Laramie Basins) and in northern New Mexico (the Moreno Valley). Bunchgrasses dominate the vegetation in these mountain flat lands. Wildflowers, although less abundant and diverse than in wet meadows, include yarrow, Indian paintbrush *(Castilleja)*, western wallflower *(Erisimum capitatum)*, and penstemon. Summer walks through these meadows (or around them, depending on how moist they are) are a delight to the senses. The vivid colors of the flowers burst forth in the sunlight; the grasses glisten with dew in the early morning. Hummingbirds, bumblebees, and butterflies come to collect nectar and pollen from the flowers. After a strenuous climb up a steep mountain trail, there is nothing like cooling your feet in a clear stream by a mountain meadow, soaking up the best of what the mountains have to offer: peace, calm, and beauty.

Lower Montane Zone

The transition from lowlands to mountain slopes in the southern Rockies involves different plant communities in different regions. For instance, around 6,700 to 7,000 feet (2,050 to 2,150 m), juniper savanna gives way to coniferous and mixed woodland in northern New Mexico and southern Colorado. This is the driest part of the southern Rockies, and this ecological transition is mostly controlled by precipitation. Farther north where climates are cooler and a bit wetter, prairie gives way to coniferous and mixed woodland as low as 5,000 feet (1,525 m). The seasonal pattern of precipitation is another important factor affecting the composition of plant communities. Regions that receive reliable moisture throughout the year have different vegetation than regions where precipitation comes only in one or two seasons of the year.

The two dominant conifers in these transitional woodlands are pinyon pine and juniper. These trees require the least available moisture of all the conifers. There are several species of pines with the common name "pinyon" (piñon in Spanish); these include Colorado pinyon *(Pinus edulis)* and border pinyon *(Pinus cembroides)*. The former species is by far the most common in the southern Rockies. Likewise, there are seven species of junipers found in pinyon–juniper woodlands in New Mexico. One-seed juniper *(Juniperus monosperma)* tends to dominate the northern part of New Mexico, including the foothills of the southern Rockies.

Pinyon–juniper woodlands range up to about 7,000 feet (2,100 m) in the more northerly parts of the southern Rockies, but warm microclimates on south-facing slopes allow this vegetation to grow as high as 9,000 feet (2,750 m) in some places. The northern limit of pinyon pine is in the Pikes Peak region of Colorado.

At lower elevations in the pinyon–juniper zone, the trees are widely spaced. In higher, moister locations, the trees grow much closer together. Pinyon requires more moisture than juniper in these woodlands, so drier regions have more juniper than pinyon and wetter regions have more pinyon than juniper. This woodland can take the heat, but it is sensitive to cold. The frost-free period associated with pinyon–juniper woodland is greater than ninety days, more than any of the coniferous forests of the Rockies. It covers more than 75,000 square miles (194,000 km^2) of the Southwest.

The general pattern of vegetation in the lower montane zone of the southern Rockies is as follows. Ridgetops and gently rolling uplands are generally forested in Ponderosa pine. The forests open up considerably on south-facing slopes. Valley floors have a rich vegetation of grasses and other herbs. Along streams, cottonwoods, willows, Ponderosa pine, and Douglas-fir grow, with occasional blue spruce. North-facing slopes support relatively dense stands of Douglas-fir, especially on steep slopes. Gentler north-facing slopes support a mixture of Douglas-fir and Ponderosa pine. Grassy openings are rare on these slopes.

At the southern tip of the Rockies, lower montane forests are generally found at elevations from 7,500 to 8,500 feet (2,300 to 2,600 m). In the central part of the southern Rockies, lower montane forests are established as low as 5,600 feet (1,700 m) on both sides of the Continental Divide. These are relatively dry forests, dominated by Ponderosa pine and, in northern New Mexico, by a mixture of Ponderosa and pinyon pines. Junipers and oaks are also common in the latter region. All species need to be well-adapted to semiarid conditions, because annual precipitation in this elevational zone averages only 18 to 25 inches (460 to 640 mm).

Douglas-fir is uniquely adapted to life on mesic mountainsides. To take advantage of the available moisture and nutrients, Douglas-fir has a *symbiotic* (mutually beneficial) relationship with certain species of soil fungi. Fungal hyphae (filamentous growths) envelope the roots of Douglas-fir and penetrate its tissues, forming structures called *mycorrhizae*. This combination of tree roots and fungal mycorrhizae greatly expands the absorption area of both the tree and the fungus, so the two species share nutrients and moisture absorbed by the mycorrhizal combination. Many other trees also have symbiotic relationships with mycorrhizae. The fungi associated with Ponderosa pine have mycorrhizae that form false truffles. These mushroom-like growths are harvested by squirrels, which helps disperse the fungal spores to new localities.

Douglas-fir stands have their own associations of shrubs and herbs. Among the shrubs found in these forests is Rocky Mountain maple, mountain-ash, and wild raspberry *(Rubus idaeus melanolasius)*. The herbs associated with Douglas-fir stands include mountain arnica, fairy slipper orchid, false Solomon's seal, and twinfower *(Linnaea borealis)*.

The southern tip of the Rockies lies in the Sangre de Cristo Range in northern New Mexico. There are precious few high mountains in this region, but there are considerable regions of montane and subalpine forest, even if there is little alpine tundra (only about 79,000 a or 31,500 ha). The foothills at elevations below about 6,700 feet (2,050 m) are clothed in either juniper savanna or plains-mesa grassland. Savanna is defined as a grassland with widely scattered low trees (in this case, juniper). The juniper species found in this type of vegetation in the southern Rockies region is one-seed juniper. High mesas (table lands) are common in northern New Mexico, and juniper savanna is the dominant vegetation of these mesas.

Foothills Zone

The foothills zone of the eastern slope of the central and northern sectors of the southern Rockies represent the transition from shortgrass prairie to lower montane forest. The prairie vegetation is adapted to extremes of climate. Mean annual precipitation ranges from 10 to 16 inches (25 to 41 cm). Hail storms, wind storms, dust storms, and tornadoes are not uncommon. Winter nighttime temperatures may drop to -20°F (-29°C) and summer daytime temperatures frequently exceed 100°F (38°C). The low precipitation keeps out trees and most shrubs, so the prairie vegetation is dominated by grasses. On the western slope, the foothills zone is highly dissected by mesas and canyons. The lower slope vegetation comprises desert scrub and desert grassland communities.

In Sum: Important Plant Communities

I am certain that you recognize that what I have written in this chapter about the botany of the Rockies is essentially a thumbnail sketch, lacking many of the details that make each region within the Rockies both unique and botanically interesting. I urge you to explore the literature cited at the end of the chapter for more in-depth treatments of individual regions and ranges and to get out and "botanize" the mountains with a field guide in one hand and a hand lens in the other. Unlike the animal life, the plants do not run away and hide. They stay put on the landscape and reveal their secrets to all who are interested in studying them.

The next section of this chapter discusses the medicinal and food uses that Native peoples have discovered for the Rocky Mountain flora. Many of these plants are not just pretty; they have properties that make them quite useful to humans.

Medicinal and Food Plants of Native Peoples

During the past 12,000 years, Native peoples who have lived in and around the Rockies developed a wide knowledge of native plants and their uses. Extracts from plants stocked the Native pharmacopeia, providing relief from ailments ranging from dandruff to dyspepsia, from carbuncles to various cancers. Until the late nineteenth century, Western (European) medicine relied almost exclusively on plant extracts as well, so there is nothing "backward" or inferior about drugs obtained from plants. A surprising number of modern medicines are still derived from plant extracts, and new and effective plant-derived drugs are being discovered every year. This is one of many good reasons to maintain the biological diversity of the planet: There are untold thousands of plant species in the tropics and elsewhere that may yet hold the cure for seemingly incurable diseases.

The Rocky Mountain Native peoples apparently shared much of their botanical knowledge with each other. Most of the uses found for plants described in this section were practically universal throughout the Rockies and western North America, wherever these plants grow. The Native healers, or medicine men, shared information with each other at gatherings. Each medicine man (or woman, in some cases), kept a large variety of dried plants and plant extracts in his or her teepee or lodge, ready to treat many different kinds of maladies.

Although we tend to think of Rocky Mountain Native peoples as hunters, they were actually hunter–gatherers who spent much of their time harvesting plant foods from the forests, meadows, and plains of the West. Game animals became scarce at times, and hunting alone could not supply the dietary needs of the people. Many wild plants were harvested in the summer and fall and then stored for winter use. Cultivated plants, such as corn, beans, and squash, were introduced into the southern Rockies region via the Anasazi culture, beginning about 2,500 years ago. Even these agriculturists continued to include many wild plants in their diet, and crop cultivation is difficult if not impossible in most of the highlands of the Rocky Mountains. So native plants were essential to the health and well-being of Native peoples in this region.

It is beyond the scope of this book to describe all the food plants or medicinal plants used by Native peoples of the region, but I offer some interesting examples and refer you to the reference section for more details.

Western Larch

Native peoples of the northern Rockies used the western larch in a variety of ways. The Flathead and Kutenai peoples made a sweet syrup from its sap by carving out a cavity in the trunks of trees, then gathering about a gallon of sap that collected there. Excess water in the sap was allowed to evaporate off, and the concentrated syrup made a tasty treat. This sap gathering was done once or twice a year, with special attention paid to trees that produced the sweetest sap.

Like subalpine fir, western larch resin (gum) was used by Native people of the northern Rockies to treat cuts and bruises. Tea boiled from the needles was used to fight tuberculosis, and the Nez Percé drank it for coughs and colds. They also chewed the sap to treat sore throats.

Ponderosa Pine

The ubiquitous Ponderosa pine had a number of uses to Native peoples. The resin was used for various medicinal purposes, including treatment of boils, carbuncles, and abscesses (the latter treatment was applied by Meriweather Lewis during the Lewis and Clark expedition, when the drug supplies ran low). Flatheads used Ponderosa pine pitch to treat rheumatism and backache. They melted the pitch over a fire and mixed it with tallow to make a poultice that was spread onto a buckskin worn on the affected area. Sweetened pitch was given as a cough syrup. Mashed inner bark was made into a poultice that was put on burns and sores.

Ponderosa pine was treasured by Native peoples throughout the Rockies for its food value. The inner bark of this tree is sweet and provides sometimes much-needed calories, especially when other plant foods are unavailable and the store of dried foods is nearly exhausted at the end of winter. Most tribes thought of pine bark as a "starvation food," to be used mostly in times of famine.

Pinyon Pine

The southwestern Pinyon pine played an important role in the diets of all the peoples who lived within its range. In the southern Rockies, these tribes included the Utes and the Pueblo, who lived along the Rio Grande in northern New Mexico. The large seeds, or pine nuts, mature in the female cones. These were harvested every autumn, as they matured. Thousands of cones could be gathered in a day by a group of people working together. Pinyon pine nuts are very nutritious. A large handful (100 g) of these seeds contains 568 calories. They are rich in protein as well as carbohydrates. Once roasted, these delicious pine nuts last through the winter without spoiling.

Willows

Willows have been used for medicinal purposes all around the world. The ancient Greeks knew of its pain-killing qualities, and willow extracts have been used in many ways in many cultures, including the Native peoples of the Rockies. Willow bark contains the compound salicyn. This is very closely related to aspirin, which is acetyl salicylic acid. Cheyennes used peachleaf willow bark shavings to make a tea to counteract diarrhea and other stomach maladies. Flatheads chewed the bark for the same ailments. Chewed bark was made into a poultice placed on cuts to speed healing and numb pain. Flatheads made an eyewash from boiled willow leaves and stem tips.

Gambel Oak

Gambel oak *(Quercus gambelii)* is the dominant oak species of much of the Rockies. Oak bark contains tannins, used as an astringent. Astringents contract or draw together the tissues, aiding in cell-wall binding after injuries such as cuts and skin abrasions. Oak galls form where insects attack twigs, causing a ball-shaped swelling of plant tissues around the source of irritation. Tannins are particularly concentrated in these galls, so Native peoples of the southern Rockies would harvest oak galls for making medicine. The gall was boiled in water to make an infusion to be applied to the skin. Tea made from oak bark was used to treat inflamed gums, diarrhea, dysentery, and hemorrhoids.

Sagebrush

The Old World name for sagebrush is wormwood. As that name suggests, sagebrush is useful in the treatment of intestinal worms. Native peoples throughout the West boiled the leaves and stems of sagebrush to make a tea for treating worms. The tea is quite bitter, but when taken every day for a period of about two weeks, it is said to be effective.

Berry Shrubs

Numerous Rocky Mountain shrubs yield berries that were an important part of the diet of regional tribes. Western serviceberry *(Amelanchier alnifolia)* was one of the most important berry producers for Native peoples in the central and northern Rockies. It was eaten when ripe, sun-dried for winter use, and pounded with dried bison meat and lard into cakes of pemmican (a dried food that could be preserved for long periods).

Chokecherry *(Prunus virginiana melanocarpa)* was another vital element in the annual Rocky Mountain berry harvest. This fruit took a bit more time to prepare, as the berries had to be smashed with a pestle and the inedible stones removed. However, having smashed the berries, Native women made them into cakes to be dried in the sun and stored for winter. They were also used in pemmican and added to winter stew pots for flavor and better nutritional balance.

Chokecherry also has several medicinal properties. Tea brewed from chokecherry bark was widely used by Rocky Mountain tribes to treat such digestive problems as diarrhea and dysentery. Lewis used a chokecherry-bark tea to cure abdominal cramps when camped on the upper Missouri River. Chokecherry bark tea was also used by Native people to rid themselves of intestinal worms. Flatheads used chokecherry bark resin as an eye medicine.

Buffaloberry *(Shepherdia canadensis)* was another common berry used by Native peoples throughout the Rockies as food. The berries were often gathered after the first freeze, when they were most sweet. The fruits were eaten raw, dried for winter, or pounded into a mash before drying. The mashed berries were often added to season buffalo meat, which is thought to be the origin of the common name.

Huckleberry is perhaps the sweetest of all Rocky Mountain berries. It is still widely harvested in the northern Rockies, and not just by Native peoples. Native peoples ate them in the late summer and autumn, or sun-dried them for use in cooking pots in winter. Unlike the other berries in this description, however, they were apparently not used in making pemmican.

Herbs

Dried herbs formed the basis of most Native medical kits, because they are lightweight and easily transported. Descriptions of some of the important herbs used to make traditional medicines in the Rockies follow.

Horsemint

Horsemint *(Monarda fistulosa),* a tall plant with a cluster of purple flowers at the top—although avoided by horses despite its name—was put to several uses by Native peoples from Alberta to Texas. Whole plants were boiled to make a weak tea taken for coughs. Dried plants ground to powder were rubbed on the skin to fight off fevers. The minty tea was also taken to sooth sore throats. Horsemint leaves, when burned in a sweat lodge fire, would increase sweating. This treatment was used for respiratory aliments such as colds and flu. One of the oils in the plant is the antiseptic thymol, the first active ingredient listed on the label of Listerine.

Osha

Osha *(Ligusticum porteri)* was used for medicine by Native peoples throughout the Rockies and western North America. The root was boiled to make a tea or ground into a powder. The plant has antibacterial properties and was used to treat coughs, prevent wounds from becoming infected, and for stomach disorders. Collecting osha root was left to the medicine men, because the plants closely resembles poison hemlock. The first mistake with hemlock is often the last one ever made. Osha grows in aspen groves up to about 7,000 feet (2,175 m).

Usnea

One of the more unusual medicinal plants of the northern Rockies is Usnea *(Usnea)*, also called old man's beard. The Dakota name for this plant translates as "the north side of the tree," which tells where to find this parasitic lichen. Usnea was used to treat diseases of the lung, intestines, throat, sinuses, and reproductive tract. It has antibacterial properties, probably from the fungal half of the lichen (fungi are the single most important sources of antibiotics, including penicillin and all of its derivatives). Ground into a powder or boiled in water to make a tea, Usnea was said to be an effective treatment for tuberculosis.

Wild Licorice

Like its European relative, North American wild licorice *(Glycyrrhiza lepidota)* was one of the essential herbs in the medicine man's pharmacopeia. The roots are the medicinal part of the plant, although some tribes also used the leaves. Licorice root tea was used to treat coughs because it counteracts muscle spasms associated with coughing and it is an expectorant. It also has antiinflammatory properties to soothe inflamed tissues. It was used to treat children's fevers. The Sioux took advantage of its antiinflammatory effects to make a poultice of the leaves and apply it to the sore backs of their horses. Licorice root tea was used to treat stomach ulcers and to fight infection. The Sioux chewed the raw root for toothaches, and they steeped the leaves in hot water and applied them to aching ears.

Licorice root was also used as a food by Native peoples in the Rockies. Lewis and Clark noted that the Native peoples in Montana roasted the root in embers, then pounded it with a stick to separate the fibrous parts from the softer edible parts. Having tried it themselves, Lewis and Clark noted that the taste reminded them of sweet potato.

Wild Roots

The roots, tubers, and other fleshy subterranean parts of plants were important foods to Native peoples of the Rockies. These foods had several qualities that made them desirable. First, they were reliable. Year after year, the plants that produced these roots would send up shoots and flowers, letting the Native people know where to find the plants and providing clues about the right time to harvest the roots. Second, they were nutritious. These fleshy below-ground parts are used by the plants to store starches, sugars, and other carbohydrates. Third, they were plentiful. The Native women returned year after year to the same meadows and hillsides to harvest roots, and would often collect enough to last their families through the winter.

Camas

Camas grows from a fleshy bulb and produces lovely blue flowers. It is also called blue camas *(Camassia quamash).* This name helps distinguish it from white camas *(Zygadenus elegans),* a plant in a different genus that superficially resembles blue camas. White camas have white flowers. The white camas is also called death camas because of its toxic effects. The blue camas bulb was one of the most important plant foods of Native people in the northern Rockies. It is mainly found in moist meadows west of the Continental Divide. Because it varied in quantity and quality from region to region, it was frequently traded among the tribes. The Shoshoni traded it to the Nez Percé; the Nez Percé traded it to the Gros Ventre and Crow. The Flatheads are said to have preferred the flavor of camas that grew in the Nez Percé's territory to that of their own.

Bands of Native people had their own camas foraging grounds that they returned to each year. Camas digging was an activity that was happily anticipated throughout the year. The ability of a young woman of marriageable age to bring a large quantity of camas back to camp was a quality not overlooked by potential suitors. Elk antlers were used as digging tools. The soil surrounding the plant was loosened with the digger until the bulb could be pulled up out of the ground. The bulbs were brought back to camp in baskets and then cooked in earth ovens before being eaten. Many regional tribes had strong taboos against men going near these cooking pits while the camas bulbs were being prepared by the women. The bulbs were often cooked with wild onion bulbs to make a tasty combination. The baking took as much as forty-eight hours.

Bulbs that were not to be eaten immediately were sun-dried for about a week. If kept dry, camas bulbs thus prepared would last indefinitely. Canadian explorer David Thompson tested the quality of long-preserved camas bulbs by offering dried bulbs to a certain Lord Metcalfe, who found them quite edible and remarked that they tasted like bread. These particular bulbs had been in dry storage for thirty-six years. The bulbs were stored in a number of ways, depending on tribal custom. Flatheads stored their bulbs on platforms they built in trees or in leather pouches hung from tree branches. Other peoples used storage pits.

Bitterroot

Bitterroot *(Lewisia rediviva, Lewisia pygmaea)* was another important food plant used by Native peoples from Colorado to British Columbia. The Bitterroot Valley and Bitterroot Mountains of Montana are named for this plant, whose scientific name, *Lewisia,* was given in honor of Lewis, who first described the plant and its use by Rocky Mountain tribes. There are two species called bitterroot, but *Lewisia rediviva* is the one that grows at lower elevations, and this is the one used for food. The Native peoples of western Montana used bitterroot extensively as a food plant. Flatheads lived in the Bitterroot Valley, and had access to the best places to find this plant. Native peoples dug the root from the ground with a specially devised tool. The

Kutenais used a three-foot-long willow stick. One end of the stick was fire-hardened, then drilled to make a hole to hold a deer antler. A shorter version of this tool held an elk antler. The plants were dug up as they bloomed. The above-ground parts of the plant were removed, and the roots were peeled and washed. Some peoples removed the inner core of the root, thought to contain the bitter flavor that gives the plant its common name. Roots not immediately eaten were dried for later use, and stored roots became less bitter with age. In the northern Rockies it was traditional for women to gather enough bitterroot to last two people through the winter. This amount was about 100 pounds (45 kg), and it took most women several days of solid digging to harvest this many roots. The roots were boiled or steamed, and berries and powdered camas bulbs were added to the pot for flavor.

Biscuit-Root

Two species of the hardy perennial biscuit-root *(Lomatium)* grow fleshy tubers rich in starch that were used as food by Native peoples in the central and northern Rockies. The tubers were dug up in early May, then peeled and eaten raw or dried for later use. Dried biscuit-root tubers were ground into flour by some people, then moistened into cakes that were baked over a fire. The Nez Percé baked biscuit-root tubers in a pit, much like camas was prepared.

Yampa

Yampa *(Perideridia gairdneri)* has a sweet-tasting root that lacks the bitterness found in so many other roots used by the Native peoples as food. Yampa roots were dug in spring or early summer. Like camas, the best time to harvest these roots is when the plant is flowering. The roots were eaten raw and also boiled or baked and made into dried cakes to preserve them for later use.

Spring Beauty

We finish our discussion of roots used as food with spring beauty *(Claytonia lanceolata),* called "Indian potatoes" by the Flathead and Kutenai tribes after they had been introduced to the domestic potato grown by white settlers. This plant figured in the annual survival of many Rocky Mountain people, because it is one of the first plants to bloom in spring, and this was a signal that the tuber-like roots were ready for harvesting. After a long winter with dwindling supplies of dried foods and perhaps scarce game, the Native peoples looked forward to seeing spring beauty blossom in the mountain forests.

Cattail

Another important wild food for Native peoples throughout the Rockies and elsewhere was cattail *(Typha latifolia).* All edible parts were available year-round, and because the plant grows at the edges of ponds, it could be found even in mid-winter when heavy snows cover the forest floor. In the spring the young shoots were pulled, the outer leaves peeled away, and the inner part eaten raw. The flower stalks, called spikes, are also edible. These were boiled, then eaten like corn on the cob. The roots were harvested in fall when the starch content is highest. These were peeled to get at the white core, which was either eaten raw or boiled, baked, or pulverized to make a flour.

Pipsissewa

Pipsissewa *(Chimaphila umbratella)* is a small evergreen shrub that has medicinal properties that were first recognized by the Native peoples. It was adopted by white settlers as they learned of its uses from the Native peoples. The Kutenai and Flatheads made a solution from this plant that they used to treat sore eyes. This solution has astringent properties that help draw fluid away from inflamed tissues. A tea made from boiling the leaves was used to treat kidney disorders. Scientific studies have shown that this medication increases urinary discharge, flushing the kidneys of infection.

This plant was used by Native peoples and white settlers alike to treat fevers. A tea was made from the plant that was taken for this purpose. The medicinal properties of this plant come mainly from a chemical compound it contains, called chimaphilin (named after the generic name of the plant, *Chimaphila).*

Yucca

Several species of yucca *(Yucca)* live in the lowlands and up to the middle elevations of the southern Rockies. The Native peoples of the Southwest used yucca for many things. The flowers were boiled for food, as were the seed pods of some species. The leathery leaves were split and woven into baskets and mats. The root was mashed to extract a soapy substance used as shampoo.

Prickly Pear Cactus

The prickly pear *(Opuntia)* is another seemingly unfriendly plant, covered with sharp spines, but it was also an important food plant to Plains peoples and to the tribes of the Rockies, where some species of prickly pear grow in the montane zone. The fruits of this type of cactus are fleshy and sweet and were eaten raw or dried for winter use. The large spiny cactus pads are difficult to deal with, and various methods were used to remove spines and get at the pulpy flesh beneath the surface. Some Native peoples split the pads from the edge, peeling back the spiny surface to expose the edible interior. Others burnt the spines off in a fire before handling the cactus pads. Seeds were added to soups or dried and then ground into a flour used in cakes.

In Sum: Medicinal and Food Plants of Native Peoples

This is just a smattering of the plants used by Native peoples in the Rockies to provide medicines, foods, or both. I do not recommend that you go out and collect these plants for food or medicine, however, unless you have an excellent knowledge of plant species. Many of the plants listed have relatives in the same genus or family that look very much like the plants discussed, but the look-alikes may or may not be safe to consume; other look-alike species are downright poisonous. The Native peoples grew up in the woods and knew their flora extremely well. Sometimes their lives depended on identifying the right plants to use for food or medicine, so they did not approach plant harvesting in a casual or haphazard way. I hope that you won't, either.

Selected References

Anderson, R. 1984. *Beartooth Country: Montana's Absaroka and Beartooth Mountains.* Helena: Montana Geographic Series.
Cannings, R., and S. Cannings. 1996. *British Columbia: A Natural History.* Vancouver: Greystone Books.
Dick-Peddie, W. A. 1993. *New Mexico Vegetation: Past, Present, and Future.* Albuquerque: University of New Mexico Press.
Hart, J. 1992. *Montana Native Plants and Early Peoples.* Helena: Montana Historical Society Press.
Knight, D. H. 1994. *Mountains and Plains. The Ecology of Wyoming Landscapes.* New Haven, Conn.: Yale University Press.
Moore, M. 1979. *Medicinal Plants of the Mountain West.* Santa Fe: Museum of New Mexico Press.
Peet, R. K. 1988. Forests of the Rocky Mountains. In M. G. Barbour and W. D. Billings, eds., *North American Terrestrial Vegetation.* New York: Cambridge University Press.
Turner, N. 1997. *Food Plants of the Interior First Peoples.* Vancouver: University of British Columbia Press.
Weber, W. A. 1976. *Rocky Mountain Flora.* 5th ed. Boulder: Colorado Associated University Press.
Zwinger, A. H., and B. E. Willard. 1972. *Land above the Trees: A Guide to American Alpine Tundra.* New York: Harper and Row.

5

Animals

The animal life of the Rockies is one of the main attractions of the region. The Rockies preserve some of the least disturbed wildlife habitats in the lower forty-eight states. People come from all around the world to view wildlife in their native habitats, especially in the national parks. This chapter covers a lot of critters, from butterflies to bears. Enjoy an armchair tour of the Rocky Mountain wildlife as you read, and then make your way to the Rockies to see them in person.

Mammals, birds, and insects are the most visible wildlife in the Rockies. For different reasons, they attract attention from tourists. Many mammals are showy, with horns or antlers, and some of them are dangerous predators. For many visitors, a visit to Rocky Mountain National Park is deemed a success if they are fortunate enough to see elk, mule deer, and bighorn sheep. Spotting a black bear or a mountain lion would be an unforgettable experience for anyone.

Birds are ubiquitous in the parts of the Rockies frequented by visitors. Some of these, like the blue bird *(Sialia)* and Steller's jay *(Cyanocitta stelleri)* are also quite showy, with beautiful blue plumage. Others, like the bald and golden eagles, are majestic birds of prey that impress with their size, strength, and hunting prowess. Some of the smaller birds, like the chickadees, entertain with their lively antics as they flit from tree to tree.

Insects are arguably the most abundant source of animal protein for terrestrial animals and a major food source for birds. Likewise many small mammals feed either mainly or exclusively on insects. Of course, the insects take their fair share of meals from mammals, often in the form of a small sip of mammalian blood.

Mammals

In the Rockies, the rodents form one of the largest and most diverse groups of mammals, ranging in size from small mice to marmots and the American beaver *(Castor canadenis).* Rodents live in a wide variety of habitats, both terrestrial and aquatic, yet none of the Rocky Mountain rodents migrates out of the mountains in winter. Instead, many species hibernate through much of the winter.

The woodchuck *(Marmota monax)* or ground hog as it is called in eastern North America, is a rodent that inhabits the northern Rockies, ranging south into northern Idaho. One sees this animal in open woodlands, as it feeds on tender, succulent plants. The woodchuck is identified by its heavy body (5 to 10 lb, 2.4 to 4.8 kg) and short legs. It digs large dens in the ground that may be 4 to 5 feet (1.2 to 1.5 m) deep and up to 30 feet (9.1 m) long. When abandoned, these burrows are used by many other species of animals. Woodchucks are a close relative of the two species of marmots *(Marmota)* that are found in the Rockies: the yellowbellied marmot *(Marmota flaviventris)* and the hoary marmot *(Marmota caligata),* which has black and white head and shoulders.

If you see a rodent scampering along in the Rockies, it is likely a ground squirrel *(Spermophilus).* As their common name implies, ground squirrels spend most of their time on the ground. Several species of ground squirrels are common in the Rockies, and some have developed the bad habit of begging for food from human visitors to the region. One of the more ubiquitous ground squirrels of the northern Rockies is Richardson's ground squirrel, which ranges from central Montana to Canada, favoring open ground such as grassland and sagebrush habitats up to about 11,000 feet (3,350 m). Richardson's ground squirrel has a smoke-gray color on its top with white or pale buff-colored belly, along with a tail that is bordered by light bands. This ground squirrel lives in open country, such as mountain meadows and grasslands, where it forages by day, eating all parts of the herbs that form its diet, including the roots. This species lives in loosely knit colonies and excavates its own individual burrow. As they forage, the ground squirrels in a group take turns sitting up on their hind legs, looking around for predators, such as foxes, coyotes, hawks, and eagles. At the first sign of trouble, the ground squirrel will give a sharp cry, and the whole colony will quickly dive into their individual burrows.

In contrast to their rodent cousins, tree squirrels are the climbers of the rodent order. Abert's squirrel *(Sciurus aberti)* and the red squirrel *(Tamiasciurus hudsonicus)* are the common tree squirrels one is likely to encounter in the Rockies. Abert's, or "the tassel-eared squirrel," is restricted to the southern Rockies, living in pine forests up to 8,000 feet (2,440 m). This squirrel is often pitch black, but it can also be salt-and-pepper gray or brown. The large tassels on its ears make it highly distinctive, so much so that people have nicknamed it the "squabbit."

The red squirrel is found throughout the Rockies, the boreal forest regions of Canada and Alaska, as well as in the eastern mixed forests and along the Appalachian Mountains as far south as the Carolinas. Hiking through the woods, one often hears this squirrel before seeing it. This squirrel, also called the pine squirrel or chickaree, produces a loud, ratchet-like call that informs intruders of its displeasure at being disturbed. It is active throughout the year, feeding on a wide variety of nuts, seeds, and fruits. It stores caches of nuts and conifer cones, which at one time were dug up by Native peoples in late winter when their own food supplies were running out. Settlers soon learned this survival skill from the Native peoples, so, in a way, red squirrels may have played a part in human history. Who knows how many hard winters were survived thanks to the caches of the red squirrel?

Although the red squirrel spends most of the daylight hours in trees, looking for food, in the northern Rockies it often nests on the ground. In the central and southern Rockies, it builds nests in conifers. The nests are made of large piles of twigs, leaves, and shredded bark. This squirrel eats a wide variety of foods found in the forest, including seeds, nuts, fungi, and eggs. Its food caches are often numerous, so if one cache is discovered by another squirrel or some other mammal or bird, it will still have other caches to see it through the winter. The home range of a red squirrel is relatively small, usually less than 650 feet (200 m) in diameter. However, each squirrel gets to know the food resources within its territory extremely well, and they defend their territories against other squirrels, especially their food caches.

Red squirrels are active throughout the year, tunneling beneath the snow in winter to find caches. One of their most important sources of food are the seeds (pine nuts) held within pine cones. Many of these cones remain on the tree through the winter months, providing much needed food for the red squirrels. The squirrels unwittingly do the pine trees a favor by burying so many cones in their caches. Not every cache is found again, or every buried cone eaten, so the squirrels serve to spread the pines' seeds some distance from the source tree.

At the high end of the rodent size scale is the American beaver. The beaver is the largest North American rodent, weighing up to 60 pounds (29 kg) and reaching total lengths of as much as 40 inches (102 cm). Few animals of any size can dictate the structure of their habitat as much as beavers. This animal is carpenter, dam-builder, hydrological engineer, and forester. Beavers create their own ponds by damming streams with tree branches, rocks, and mud. Most of the beaver dams are about 6 feet high and 200 feet long (about 2 m by 61 m), but the largest beaver dam on record was 13.7 feet tall and 3,280 feet long (4.2 m by 1,000 m). Once beavers build their dam and the stream backs up to fill a basin, creating a pond, the beavers proceed to fell trees to supply their food through the winter, sitting cozily in their lodges made of large mounds of tree branches and twigs. The only entrance to a beaver lodge is well below the waterline of the pond, making it virtually impossible for predators to attack beavers in their lodges. The beaver has a large, flat tail, shaped like a paddle. It uses this tail to help swim and slaps the surface of the water to warn its family of an approaching predator.

Beavers were once plentiful throughout western North America, and it was the lure of beaver pelts that brought the first round of European explorers to the Rockies. Beaver fur was used to make men's felt hats at the beginning of the nineteenth century. Within fifty years, nearly all of the beaver in the Rocky Mountain region had been trapped out. Perhaps the only thing that saved the beaver from total annihilation was a change in fashions, dictating that hats should be made of silk rather than of beaver fur.

The absence of beavers in the Rocky Mountains for nearly a century (roughly 1830 to 1930) brought changes in many mountain drainages that can still be seen today. Beaver ponds slow the runoff of stream waters. Without them, streams tend to flood more often and more spectacularly. Old flash-flood channels from the non-beaver period can still be seen in some regions. By creating ponds, beavers also serve as agents of biodiversity. They create aquatic and wetland habitats for fish, frogs, ducks and geese, shore birds, and muskrats.

Beaver ponds are not maintained indefinitely by beavers. In fact, as soon as the dam is built, the pond begins to accumulate sediment from upstream waters. Eventually, the pond fills in with sediment, and beavers begin again elsewhere. In the meantime, the in-filled pond becomes a bog that is colonized first by herbs and forbs, then shrubs such as willow, alder, and birch, and finally by trees from the surrounding forests. This whole process may take thousands of years to complete, but the ecological diversity of the beaver pond locality is greatly enhanced by the creation of wetland. Would that all dam builders and engineers were as useful to local ecosystems as beavers.

One of the most interesting rodents of the Rockies is the bushytail woodrat *(Neotoma cinerea)*. This inhabitant of pine forests and rocky slopes weighs between 11 and 12 ounces (300 and 350 g) and has a soft coat of golden to buff-colored fur on top, white fur underneath, and a bushy tail (especially adult males). The other common name for the woodrat is packrat. The bushytail woodrat has a deep sense of curiosity about its surroundings. On its nightly foraging trips, it investigates everything in its path, and if an object catches its fancy, it takes it home to its den to add to its collection of curios. Since coming into contact with people, woodrats have begun collecting such items as spoons, watches, aluminum foil, corn cobs, false teeth, and diamond rings left by the kitchen sink. But, as one researcher mused, the bushytail woodrat is a packrat with a conscience. When it takes something from you, it generally leaves something in exchange. What really happens is that the woodrat ambles along with several items in its mouth, discovers something interesting, and has to abandon one item to make room in its mouth for a new one. Finley (1958) provided a list of items found in bushytail woodrat nests in Colorado:

- tar paper
- nails
- rattlesnake carcass
- peach pit
- rope
- leather glove
- shotgun shell
- lump of coal
- bolt and spare nut
- hacksaw blade
- wire
- porcelain insulator
- sticks
- stones
- dung of other animals

The bushytail woodrat feeds mainly on plants, and it will eat almost any kind of plant, including species deemed noxious by other herbivores because of the toxic secondary compounds in their tissues. Because of their adaptability, this species ranges throughout the Rockies, pilfering interesting natural and humanmade objects from Santa Fe to Jasper.

Another unusual rodent resident of the Rockies is the porcupine *(Erethizon dorsatum)*. Porcupines use their large incisors to gnaw the bark of trees. The inner bark

COREL CORPORATION

Marmot

COREL CORPORATION

Bluebell

Star gentian

COREL CORPORATION

Grizzly bear with cubs

COREL CORPORATION

Coyote

Bison

JAMES HALFPENNY

Yellowstone wolf

COREL CORPORATION

COREL CORPORATION

Mountain lion

COREL CORPORATION

Pronghorn antelope

Canadian anemone

COREL CORPORATION

Bunchberry

Wild rose

COREL CORPORATION

Yellow monkey flower

COREL CORPORATION

Spring beauty

COREL CORPORATION

Fireweed

Bald eagle

Elk cow and calf

Mountain goat

Bighorn rams

Great horned owl

COREL CORPORATION

COREL CORPORATION

Woodrat

COREL CORPORATION

Beaver

Pipevine swallowtail butterfly

COREL CORPORATION

Buckeye butterfly

COREL CORPORATION

Callipe fritillary butterfly

COREL CORPORATION

COREL CORPORATION

Greenish-blue butterfly

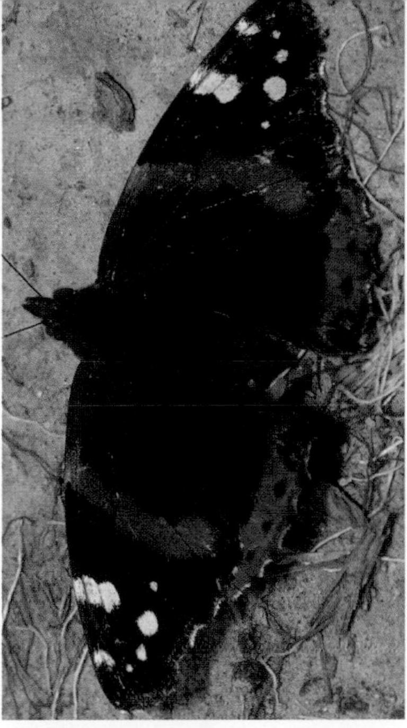

COREL CORPORATION

Red admiral butterfly

COREL CORPORATION

Viceroy butterfly

Cliff Palace, Mesa Verde National Park, Colorado

WILLIAM KRANTZ

Patterned ground, Medicine Bow Mountains, Wyoming

Meandering stream, Rocky Mountain National Park, Colorado

COREL CORPORATION

Pacific Northwest forest, Kootenya National Park, British Columbia

Mosses on forest floor, Pacific Northwest forest, Glacier National Park, Montana

Moss campion

Lower Falls, Grand Canyon of the Yellowstone, Yellowstone National Park, Wyoming

Perched glacier on side of mountain, Icefields Parkway, Alberta

Grass-of-Parnassus

Fossil corals in rock from Parker Trail, Icefields Parkway, Alberta

Mount Athabaska, Alberta

Sky pilot

Mammoth Hot Springs, Yellowstone National Park, Wyoming

Geyser basin 1, Yellowstone National Park, Wyoming

Hot springs and terraces 1, Yellowstone National Park, Wyoming

Hot springs and terraces 2, Yellowstone National Park, Wyoming

Old Faithful Geyser, Yellowstone National Park, Wyoming

Geyser basin 2, Yellowstone National Park, Wyoming

River beauty

Purple saxifrage

COREL CORPORATION

Midge (fly)

is particularly delectable to the porcupine, because it contains the most nutrients. Porcupines also feed on buds and small twigs; in spite of their slow, clumsy gait on land, they are good climbers, and frequently sleep near the tops of trees. Of course the most unique (and potentially painful) attribute of the porcupine is its quills. Porcupine quills are modified hairs. The quills are from 2 to 3 inches (6 to 8 cm) long. Their tips are barbed, making their extraction from a would-be predator's hide a painful process. In fact the quills tend to work their way deeper and deeper into the flesh, and may eventually cause the death of a victim. The porcupine's belly is its only vulnerable area, because it has no quills. If a coyote, mountain lion, or grizzly is to attack a porcupine without getting a face full of quills, it must first find a way of flipping the porcupine onto its back. Thus the porcupine tends to stand its ground in the face of predators, a strategy that works extremely well against other mammals but extremely poorly against the wheels of speeding cars. That is why one sees so many road-killed porcupines in the Rockies.

Rabbits and hares (Lagomorpha), or lagomorphs as scientists refer to them, are common in some parts of the Rockies. The snowshoe hare is the quintessential Rocky Mountain rabbit, superbly adapted to life in snowy habitats. As its name implies, this hare has extra-large feet, which it uses like humans use snowshoes, to distribute its weight over a large region of snow so that it will not sink into the snowpack. Most hares and rabbits have large ears, but the snowshoe hare has very small ears, thereby reducing exposure of these blood-rich extremities to cold winter air. Snowshoe hares are brown in summer and turn white in winter, the same camou-

COREL CORPORATION

Snowshoe hare. This species is common throughout the high country of the Rockies.

flage technique used by ptarmigan *(Lagopus)*. Snowshoe hares range throughout the boreal and much of the arctic regions of North America and throughout the Rockies. In the southern Rockies, snowshoe hares can live as high as 11,000 feet (3,350 m). In summer they feed on grasses and forbs, but in winter they turn to the bark and young twigs of shrubs and other woody plants. They are most active at night, and by day seek shelter in depressions under shrubs or other cover. Snowshoe hare populations are known to go through an eleven-year cycle between boom and bust, although the reasons for this cycle remain unclear.

There are five species of deer in the Rockies: moose *(Alces alces)*, elk *(Cervus elaphus)*, woodland caribou *(Rangifer caribou)*, mule deer *(Odocoileus hemionus)*, and whitetail deer *(Odocoileus virginianus)*. The moose is the largest of these five species, weighing as much as 1,180 pounds (560 kg) and standing up to 7 feet (2.1 m) tall at the shoulder. This large, gangly animal has been said to look like it was designed by a committee. It has a large, overhanging snout, a pendulous "bell" hanging from its throat, a torso resembling that of a domestic cow, and the legs of a horse. The bulls have massive, flattened antlers with projecting prongs. These antlers are quite heavy and can be quite large. The largest recorded moose antlers spanned more than 77 inches (196 cm). The antlers are shed in winter, with a new set of antlers grown each year. This species ranges across the boreal region of North America, and native populations extend along the Rockies as far south as Wyoming. Moose were introduced into North Park, Colorado, beginning in 1978. They have since begun wandering into Rocky Mountain National Park and are now well-established in the Colorado River drainage of the park. This is somewhat troublesome to park wildlife managers, because the moose is not truly native to the park, so it must be considered an alien species. Park officials, however, are not trying to eradicate moose from the park.

Moose spend much of their time in or near standing water. This large deer is a good swimmer. Burt (1964) noted that in open water, moose can swim as fast as two men can paddle a canoe. This observation sounds as if the author conducted some field testing, though perhaps not willingly. Moose are especially fond of aquatic vegetation, and are often seen plunging their heads into the water, coming up with a mouthful of plants torn from the mud at the bottom of the lake. Most of their foraging activity is at night, but they can be active any time of day. In winter, when ponds and lakes are frozen over, moose browse on the twigs and bark of deciduous trees and shrubs, as well as saplings. Because of their northerly distribution, moose have never been threatened with extirpation from the major parts of their range.

Elk, or wapiti, are also large animals, though not as large as moose. Adult bull elk weigh from 800 to 1,000 pounds (380 to 475 kg) and the cows generally weigh 500 to 600 pounds (240 to 280 kg). Bull elk grow impressive antlers, with a beam length of about 64 inches (163 cm). Like moose, elk shed their antlers in winter and must grow new ones in spring and summer. Elk range today throughout the Rockies, although they were extirpated from the Canadian Rockies near the turn of the twentieth century and had to be reintroduced from U.S. populations. Elk herds had also declined dramatically in the southern and central Rockies in the early part of the twentieth century, mainly from overhunting by market hunters who were supplying meat to the Rocky Mountain mining camps and Denver hotels. In contrast to the market hunters, the sport hunters of the past century and into the twenty-first are in large part responsible for the reestablishment of elk herds in the Rockies. Elk numbers have risen to high levels, perhaps all-time highs in some regions, and are thus fairly easy to observe, especially in the national parks.

Elk are herd animals, traveling in groups of thirty to fifty individuals. One can commonly see elk herds grazing in mountain meadows or dozing in the shade of tall trees. The herds tend to migrate to higher elevations in spring, and often congregate on the alpine tundra in summer, where they feed on tundra herbs. In fall, elk move back down the mountains, where they winter in montane forests. Their winter food consists mainly of tender branches and the bark of shrubs and trees. In many regions, it is possible to discern the depth of a previous winter's snow by observing the height to which the bark has been stripped from aspen trees. Indeed, heavy elk consumption of aspen bark is killing many trees and threatening whole stands of aspen, especially in national parks of the Rockies where elk populations have soared in recent years.

During late summer and early fall, male elk become more aggressive with each other, and the older bulls begin herding a group of cows into a mating herd, or harem. This is a full-time job, complete with courtship rituals, displays of male aggressiveness, and loud, high-pitched calls (bugling) that demonstrate the presence of bulls to prospective cows and warn other bulls away. The bugling and demonstrations of dominance by assorted old bulls tends to encourage the cows to wander from one harem to another. The most senior bulls can usually manage to keep several cows to themselves, but they refrain from eating for days at a time to keep their harems intact.

The mature females (cows) become pregnant in the autumn, and gestation lasts eight and a half months. The young are born from mid-May through early July, depending on where the elk live. Calving time generally coincides with the period when the most nutritious food is available. When a cow is ready to give birth, she often leaves the herd and seeks a site that elk seldom visit. This strategy helps her avoid predators that are familiar with traditional elk migration and grazing patterns. A newborn calf is slow to get up on its feet and cannot run fast or for very long. Instead, camouflage (its spotted coat) keeps it safe from most predators. The cow also helps hide the calf by eliminating any signs of her calf that a predator might detect, such as the afterbirth (placenta) and the calf's feces. To avoid leading predators to her young, she grazes away from the hidden calf and returns to nurse it only a few times a day. If a predator approaches, though, she will attack or try to lead the animal away from the calf. Calves have the summer to grow and store up energy for the hard times of winter. One calf per birth is normal, and two to three calves per birth is rare (< 1 percent). Calves weigh 20 to 45 pounds (9.5 to 21 kg) at birth; the average weight is 30 pounds (14 kg).

Once a newborn elk calf gets to its feet, it is able to walk and follow its mother in three days and begins grazing in four weeks. Once a calf can run, the cow and her calf rejoin the herd. This takes one to three weeks. Calves are weaned in late summer but follow their mothers until the following spring. Once with the herd, calves are grouped together and watched over by a few cows. This frees the other cows to graze. Pregnant cows often get a head start on the annual migration to the high country by beginning their migration before they give birth to their calves. They may return to the same range every year, using similar routes and river crossings. In the Rockies, they resume moving into higher country by late June or July, where they will find rich summer food.

One of the first impressions easterners have when they see a deer in the Rockies is "Wow, the deer are different here!" The reason is that the most common deer of the montane forests of the Rockies is the mule deer, named for its large, mule-like ears. Mule deer range throughout the Rockies and most of western North America south of the arctic tundra. This species is considerably smaller than elk or moose, standing only about 3 feet (91 cm) tall at the shoulders. Adult mule deer weigh as much as 400 pounds (190 kg). The antlers of mule deer bucks are branched into equal Y-shaped parts, which differentiate them from whitetail deer, whose antlers simply have tines projecting from a single main stem. Mule deer do not raise their black-tipped tails when running, in contrast to whitetail deer, which often raise their white tails. Both species are browsers, feeding on the leaves and new shoots of shrubs in summer and on bark and small stems in winter. Shrub leaves represent about 75 percent of mule deer summer diet, and the leaves of herbaceous plants make up the rest of their summer diet. Mule deer are currently quite abundant in most regions of western North America, although only a century ago this deer had been hunted out of many parts of the Rockies.

Mule deer are important prey animals for mountain lions and wolves. The young are especially vulnerable to predation. Coyotes, lions, and bears prey on young fawns in the southern Rockies (where wolves are virtually extinct). A study of sixty-four young fawns in the Rockies of Colorado showed that 25 percent died from attacks by these predators. Even grizzly bears, the largest predators in the Rocky Mountains, rarely attack an adult mule deer, unless it is weakened by extreme old age or disease. Adult mule deer have sharp hooves and antler tines, both of which they use to defend themselves against predators. Although a mountain lion or wolf would probably eventually dispatch a healthy, adult mule deer, the risk of serious injury is just too great.

The pronghorn, often called an antelope, is neither closely related to the Old World antelopes nor the deer but rather is in a family of its own. Surprisingly, it is more closely related to bison *(Bison bison)* than to any other animal. The pronghorn has true horns, which are kept year-round (as opposed to antlers, which are shed and must be grown anew each year). However, the outer sheaths of these horns are shed each year, even though the horn cores remain intact. Both sexes have horns that have a single prong extending forward on the head. Pronghorn is more diminutive than any of the deer of the Rockies. Adults weigh from 75 to 130 pounds (35 to 62 kg) and stand about 3 feet (90 cm) tall. Though the pronghorn is not closely related to African antelope, they share an important characteristic: blazing speed and agility. Burt (1964) stated that pronghorn reach speeds of 40 miles (65 km) per hour, but I have observed a small herd of pronghorn keeping up with a car going 50 miles (80 km) per hour for a considerable distance, and then speeding up faster than this to cross in front of the car, going perhaps 60 miles (100 km) per hour in a burst of speed.

Pronghorn range from Mexico to southern Canada. This is a species of open country, and its presence in the Rockies is limited to broad valleys and more arid regions that lack dense forests. It is active by day, and is a grazer. It has a much smaller stomach than any of the deer or bison, so it seeks out the most nutritious plants for food. Pronghorn numbers declined to fewer than 30,000 animals because of overhunting in the 1800s. Now their numbers have recovered remarkably well, and today the herds total about twenty million animals.

Pronghorn are preyed on by coyotes, bobcats, and grizzly bears. The young are sometimes taken by golden eagles. I once saw a remarkable scene in the Rockies of northern New Mexico: A small herd of pronghorns (perhaps ten to fifteen individuals) were surrounded by a group of six coyotes, who were running them first in one direction, then in another, looking for young or feeble individuals. This scene was so remarkable because this style of hunting is more commonly associated with wolves. Coyotes generally hunt alone, not in packs. The chase went on for fifteen or twenty minutes as I watched, spellbound, from a promontory about 100 yards away. The pronghorns never broke ranks, and if there were any vulnerable individuals within their herd, they remained well-concealed. Eventually, the coyotes called off their hunt and loped away. Predators are only willing to expend so much energy on a hunting expedition. If they chase too long and run too far, they end up burning more calories than their potential meal has to offer.

An unmistakable site in the Rockies is a viewing of a group of mountain goats *(Oreamnos americanus)*. Mountain goats stand about 3 to 3½ feet (90 to 107 cm) tall and weigh from 100 to 130 pounds (48 to 62 kg). They have long, white fur and short, smooth, black horns. These animals are amazing climbers, beautifully adapted to their native habitat, which is steep, rocky slopes near treeline, and sometimes along roadside cliffs at higher elevations. They are primarily active in daytime, because they have little to fear from predators, few of which can climb up to where the mountain goats live. Their lives are not without danger, however. Occasionally a golden eagle will succeed in knocking a kid off its feet or in snatching one from the face of a cliff, and avalanches and rock slides have killed many mountain goats.

The mountain goat is part grazer and part browser. It feeds on alpine grasses, mosses, lichens, and various woody plants and herbs. Mountain goats are capable of more dangerous climbing maneuvers than most skilled human mountain climbers. These goats have been seen leaping across deep chasms. One goat that was seen walking along a ledge above a deep canyon got stranded when the ledge narrowed to nothing. The goat then did a cartwheel to turn around, planting its front feet on the ledge and walking its back legs up along the cliff face until it got turned around.

A mountain goat's feet are superbly designed for climbing and holding onto steep rocky surfaces. The two-toed hoof has a stiff outer rim for getting a foothold, and the inner part of the hoof has a flexible pad for getting a good grip. The dew claws help hold the goat back as it climbs downslope. There is often little for a mountain goat to eat in its rocky domain, however, high above the more fertile valleys and gentle slopes where most of the other large grazers live. Thus it will fight to protect its feeding grounds, and tends to lead a more solitary life than many other grazers.

Often mistaken for mountain goats at a distance, bighorn sheep *(Ovis canadensis)* are another great climber and symbol of the Rockies. This sheep is about the same height as the mountain goat, but they weigh as much as 275 pounds (130 kg). Bighorn sheep range throughout the Rocky Mountains and extend into the intermountain region west of the Rockies. They have dusky-brown fur with a creamy white belly. The males have massive, curled horns and the females have much smaller, straight horns. Like the mountain goat, the bighorn sheep is both a browser and a grazer, but unlike the mountain goat, the bighorn sheep moves downslope to lower mountain elevations in winter. Also, mountain goats stick more closely to their rock-ledge habitat, whereas bighorn sheep frequently come down off their lofty perches to feed in mountain meadows. These sheep are gregarious. Bands are usually made up of groups of ewes and their offspring or groups of rams that are reaching sexual maturity, led by a few older rams. Once the spring begins, the older males split off and remain more solitary until the autumn, when the annual rutting season brings the old rams together for ritual combat. This is where the big, heavy horns come in. The rams butt heads, striking blows that would probably kill other animals of this size that were not endowed with the helmet-like hardware on their heads. These tests of strength and endurance sometimes end in the death of the weaker individual, but usually the dominant male succeeds in humbling his opponent, who then retreats and waits for another chance to rise in the sheep-mating hierarchy.

Although one might confuse mountain goats and bighorn sheep at a distance, there is no confusing the bison, or American buffalo. The term *bison* is much preferred over the term *buffalo* by mammalogists, because this animal is only distantly related to true buffalo of Africa and Asia. The largest land mammal in North America, bison stand up to 6 feet tall (1.8 m) and weigh up to 2,000 pounds (950 kg). The massive head and forequarters (including a hump on the shoulders) of this species appear out of proportion to its slim hindquarters.

Once the dominant grazer of the Great Plains, bison also lived in open woodlands and forest clearings. Bison's prehistoric range included much of North America, and its numbers are estimated to have been between forty and sixty million in prehistoric times. Enormous herds roamed the plains, and much smaller herds are thought to have migrated through the Rockies. We do not know much about Rocky Mountain populations, because most mountain bison were exterminated before scientists had an opportunity to study them. Early historic records show that bison lived in mountain parklands in Colorado and in the Yellowstone region of Wyoming. Seventeenth- or eighteenth-century bison remains have been found as high as 11,600 feet (3,535 m) in the Colorado Front Range, and Native peoples such as the Utes hunted them in the Rockies during the summer months.

Bison are grazers, and like domesticated cattle, they are ruminants. Ruminants overcome their digestive obstacles of low-nutrition, high-cellulose diets by having a four-chambered gut. One of these chambers is the rumen, where digestive enzymes and microorganisms go to work on ingested grasses. To break up the plant tissues more completely, following the initial attack by digestive enzymes, bison and other ruminants regurgitate the contents of the rumen into their mouths. This "cud chewing" helps ruminant grazers to crop grasses at one time and digest them later. The rumen is a fermentation chamber, where microorganisms can break down cellulose into starches and sugars that the bison can then digest for themselves.

Bison can seem like docile animals—much like an overgrown cow—but in reality they can be extremely dangerous. Bison injure more visitors to Yellowstone Park than any other animals. When people get within 300 feet (100 m) of the bison, the animal may suddenly charge, galloping at 40 miles (65 km) per hour, then goring the fleeing human with its sharp horns. The human intruders are often propelled several feet into the air. Needless to say, the return to Earth after such a catapult is often accompanied by broken bones.

The millions of bison that dominated the plains were reduced to fewer than 1,000 individuals by the early part of the 1900s. But regional wildlife managers and private land owners (including Ted Turner of CNN fame) have built up herds of bison in recent decades. Populations have now recovered to a less dangerous level, with about 200,000 individuals in the United States and Canada. The bison would probably have been completely exterminated were it not for a public outcry toward the end of the nineteenth century, led by such individuals as William T. Hornaday and J. A. Allen.

Healthy adult bison are too dangerous to be attacked by even the largest predators of North America. Their horns and hooves are formidable defensive weapons. However, grizzlies, wolves, mountain lions, and occasionally black bears attack young bison calves, and wolves, coyotes, and bears feed on bison carrion, an important source of wintertime food for these predators.

Wolves and other carnivores instill a mixture of admiration and trepidation among humans. Of all the carnivores in the Rockies, humans need only fear three species: the black bear, the grizzly bear, and the mountain lion. The rest are either too small to attack people (such as the weasels) or are too wary of human contact to pose a real threat (such as the coyote and gray wolf).

The black bear is not a pure carnivore but rather an omnivore, eating considerable amounts of plant foods, especially berries, nuts, and roots. It also eats insects and scavenges carrion. Black bears are not great hunters and rarely take on large animals as prey. The black bear is generally shy of people and is only dangerous when surprised, especially sows with cubs. Nevertheless, they are strong, powerful animals that need to be treated with great respect. Adult males can weigh up to 500 pounds (225 kg) and females weigh up to about 150 pounds (70 kg), The name "black bear" can be misleading, because the coloration of this bear ranges from cinnamon brown to black, and there is even a white race of black bears in British Columbia.

In prehistoric times, black bears ranged over most of North America, as far north as the edge of the Arctic. This species seems always to have preferred forested regions, and was probably never common in open country. Now its range is greatly diminished, but there are still healthy populations in the Rockies, especially in the more rugged and remote ranges of the central and northern Rockies.

Many visitors to Yellowstone National Park and other parks in the Rockies who came before the 1970s remember black bears begging for food by roadsides. These bears lost their fear of humans while developing a taste for human food. As a result, they became far more dangerous to humans than black bears living in more remote regions who had no regular contact with humans. Statistics on bear attacks in Yellowstone show that the number of black bear attacks decreased markedly when the National Park Service moved the bears away from the roads and strictly enforced the "Don't feed the bears" rule. In fact, black bear attacks on humans have dropped close to zero since Park Service policy has changed.

The other bear native to the Rockies is the grizzly bear *(Ursus arctos)*. The name *grizzly* refers to the grizzled (meaning "white- or blond-tipped") hair that covers much of the upper surface of this bear. Although omnivorous, grizzlies do more hunting than black bears. Grizzlies regularly kill large-game animals such as moose, deer, and elk, preying mainly on calves or old, weak individuals. Grizzlies are also powerful diggers, and they spend much of their time excavating rodent burrows in search of prey. It is not difficult to distinguish between the grizzly and black bear.

The grizzly is larger, weighing up to 850 pounds (400 kg) and has a distinctive hump at the shoulder. If a bear chases you through the woods and you climb a tree to escape, watch for these signs: If the bear climbs up after you, it is a black bear; if it knocks the tree over to get at you, it is a grizzly.

Grizzlies roam widely in search of food, and they are dangerous enough to pose a real threat to the safety of people and their livestock. Unlike black bear attacks, which have declined markedly since park rangers moved roadside black bears into the back country, grizzly bear attacks, although rare, continue to go up and down year by year. Of the millions of visitors to grizzly country, very few have problematic bear encounters. Indeed, your chances of being attacked by a grizzly are extremely remote. A little common sense goes a long way to avoiding this kind of trouble.

Most grizzly encounters take place on back-country trails, when hikers surprise bears (especially sows with cubs). This problem can usually be avoided if hikers take a few simple precautions, such as talking while hiking or wearing "bear bells" that jingle as the hiker walks. These let bears know that hikers are nearby. Nearly all bears (black and grizzly) will shy away from humans when they know they are coming. Surprising a bear is what gets its defensive instincts aroused.

Before the 1800s, grizzly bears inhabited much of the Great Plains as well as the mountains of western North America. One of the first well-documented encounters with grizzlies came in 1805, when the members of the Lewis and Clark expedition met up with these bears in southern Montana. Their first major encounter took place on May 4 of that year. Clark described it this way in his journal:

> We saw a Brown or Grisley beare on a sand beech, I went out with Geo Drewyer & Killed the bear, which was very large and a turrible looking animal, which we found verry hard to kill we Shot ten Balls into him before we killed him, & 5 of those Balls through his lights. This animal is the largest of the carnivorous kind I ever saw.

On June 14, Lewis had a more frightening grizzly bear encounter near the site of Great Falls, Montana. He had just shot a bison and was standing over his kill when the bear approached. His journal entry for that day includes the following:

> . . . and having forgotten to reload my rifle, a large white, or rather brown bear, had perceived and crept on me within 20 steps before I discovered him; In the first moment I drew my gun up to shoot, but at the same instant recolected that she was not loaded and that he was too near for me to hope to perform this operation before he reached me, as he was then briskly advancing on me; it was an open level plain, not a bush within miles or a tree within less than three hundred yards of me; the river bank was sloping and not more than three feet above the level of the water; in short there was no place by means of which I could conceal myself from this monster untill I could charge my rifle; in this situation I thought of retreating in a brisk walk as fast as he was advancing untill I could reach a tree about 300 yards below me, bit I had no sooner terned myself about but he pitched at me, open mouthed and full speed, I ran about 80 yards and found he gained on me fast, I then run into the water the idea struck me to get into the water to such a debth that I could stand and he would

> be obliged to swim, and that I could in that situation defend myself with my espontoon [a weapon?]; accordingly I ran haistily into the water about waist deep, and faced and presented the point of my espontoon, at this instant he arrived at the edge of the water within about 20 feet of me; the moment I put myself in this attitude of defence he sudonly wheeled about as if frightened, declined to combat on such unequal grounds, and retreated with quite as great precipitation as he had just before pursued me.

Lewis later added, "I felt myself not a little gratifyed that he had declined the combat."

Since that time, the grizzly bear has been systematically hunted and has been eliminated from most of its prehistoric range. Within a century of the Lewis and Clark encounters, the grizzly bear had been wiped out in both Colorado and New Mexico. It still remains in the Yellowstone region of Wyoming, in Glacier National Park and other mountain wilderness areas of Montana and Idaho, and in the Rockies of Alberta and British Columbia, as well as in the Arctic, where it is just as much at home as it is in the Rockies.

The members of the dog family found in the Rockies include the coyote *(Canis latrans),* gray wolf, and three species of foxes. The prehistoric range of the gray wolf included all of the Rocky Mountain region, as well as much of the rest of North America outside of the southeastern United States. But wolves were systematically shot, trapped, and poisoned out of much of their original range during the past 300 years. Until recently, the only remaining Rocky Mountain populations of wolves have been in the Canadian Rockies and the most remote parts of the Rockies in Montana and Idaho. Wolves have recently been restored to Yellowstone and central Idaho.

Wolves are the largest North American canid, weighing up to 120 pounds (57 kg) and standing as tall as 28 inches (71 cm) at the shoulder. Wolves are social animals, running in extended family groups, or packs, that usually number ten to twelve but can reach as many as twenty. Wolves hunt caribou, moose, elk, deer, bison, and bighorn sheep, but they also take much smaller animals, such as hares, rabbits, and rodents. They are highly territorial animals, with hunting territories of about 60 square miles (150 km^2), but the size of the territory varies considerably, depending on food resources. In regions with few prey animals, such as the high arctic, wolf pack territories are as large as 5,000 square miles (12,500 km^2). Female wolves come into breeding condition only once a year, in winter. Once the pups are born, the mother nurses them in her den for thirty-four to fifty-one days. She rejoins a pack that usually consists of the alpha pair, its young of the year, and other (nonbreeding) adults. When the two-month-old pups emerge from the den, the adults move them to a meadow or open area near cover. This is the rendezvous site to which the adults return after a kill, bringing food to the growing pups.

When they are out hunting, the wolf pack searches for prey, calling to one another by barks and howls. Imagine that a small herd of bison is detected, grazing by a river. The wolves fan out around the bison, circling the herd, which is soon alerted to their presence. Thus the age-old contest begins again. The wolves begin to chase, trying to get the bison to abandon their defensive circle and start running. When the bison herd is running, it gives the wolves an opportunity to attack stragglers, either young ones that get separated from their mothers or old ones that are lame or otherwise infirm. The pack spots a sickly bison that cannot keep up with the herd, and several wolves move in for the kill. The first step is to bring the bison down to the ground. This time the wolves are fortunate: The old bison becomes exhausted and falls to the ground. Two of the wolves make sure that the bison's hind end stays down by clawing at its rump. Another wolf clamps his jaws over the snout of the prey, suffocating it. In a few minutes the bison stops struggling.

The pack regroups around the kill, but the younger wolves keep their distance while the dominant (alpha) male satisfies his hunger. The social hierarchy of the pack dictates the order of feeding on the carcass. The only exception is that the young wolf pups are allowed in before their "turn." Were this not the case, the youngest wolves, lowest in the social hierarchy, would not get the nutrition they need to grow and mature. The pack cannot consume all of the bison meat in one meal, so they drag the carcass into the woods and bury it with leaves, twigs, and dirt. They will come back in the evening or the next day to eat the rest.

The only wild animal likely to be confused for a wolf is the coyote. Coyotes are markedly smaller than wolves, weighing 50 pounds (24 kg) or less. Coyotes run with their tails held down, whereas wolves tend to run with their tails held out behind their bodies. Because one often sees these animals on the move, this trait is a good clue to their identity. Coyote fortunes improved somewhat in North America during the twentieth century. As the wolf was extirpated from most regions, coyote numbers increased. Wolves are known to kill coyotes that try to inhabit the same territory. Coyotes do much better than wolves in close proximity to humans, and are frequently seen in and around towns and even large cities. Coyotes are chiefly nocturnal, and they are excellent scavengers. They hunt for rodents, rabbits, and other small animals.

Coyotes have excellent hearing and sense of smell, which they put to good use when hunting small mammals in deep snow. Rodents such as voles or deer mice may be active beneath the snow, running along the ground looking for food. Amazingly, coyotes can hear these sounds through several feet of snow. When they detect a prey animal beneath the snow pack, they arch their back, leap up, and plunge their snout into the snow, snatching the rodent from its seemingly safe tunnel beneath the snow. This method does not always work, but if it succeeds three or four times per day, the coyote has obtained enough calories to survive.

Three species of cats round out our list of Rocky Mountain predators. Cats (family Felidae) that inhabit the Rockies include the lynx *(Lynx canadensis),* bobcat *(Lynx rufus),* and mountain lion *(Felis concolor).* All three cats are solitary hunters, capable of roving many miles per day in search of prey.

The mountain lion is the largest of the feline predators of the Rockies and the most closely associated with the mountain chain. This cat has made a big impression on Americans since the first Europeans landed on the east coast. The residents of many regions have their own nickname for the mountain lion, and in fact it probably has more names than any other North American mammal. Among these are cougar, puma, panther, and painter. Mountain lions once ranged throughout much of North and South America, but like the other large predators the mountain lion has been eliminated from most of its former range and now lives only in the more remote parts of the continent. This cat ranges far and wide in search of prey animals, such as mule deer. The home range of the mountain lion can be up to several hundred square miles, a factor that predisposes the mountain lion to run afoul of ranchers, farmers, and city dwellers. Western state governments have historically considered mountain lions as vermin, which means that there was no limit on the number that could be shot or the time of year they could be hunted. Attitudes toward mountain lions and some state laws have changed in the past twenty years, but even now, with their numbers very low, more than 1,800 mountain lions are shot every year in the western United States by licensed hunters.

Another 100 to 200 are shot by the U.S. Department of Agriculture, which responds to complaints by ranchers and farmers when their livestock have been killed. Bounty hunting of mountain lions took place for almost 300 years. In the 1800s the U.S. government employed as many as 200 bounty hunters per year to hunt cougars, "for the sake of the deer population." Mountain lion researchers generally consider it a misconception that the lion is responsible for any decline in deer herds. Reduction in deer herds is more likely attributable to the impacts of habitat loss and hunting.

The mountain lions' method of hunting is stealthy. Like other large cats, it creeps up on unsuspecting deer or other prey, pouncing at the last second so that it minimizes the chase. If a mountain lion can surprise a prey animal, it minimizes the risk of being kicked or gored by a struggling ungulate. Predators have to be in top shape to catch large prey animals, and anything that reduces their vigor diminishes their hunting success.

In summer, mountain lions follow the migrating deer and elk herds to higher ground in the Rockies. The lions' home range expands accordingly. In winter the prey animals move down into valleys. As their ranges contract, so does the mountain lion's. These days there are increasing numbers of mountain lion "incidents" in and near towns that lie in the foothills along either side of the Rocky Mountains. The increase in encounters is probably related to deer coming into towns to feed, especially in winter. A backyard garden full of ornamental shrubs and other plantings is like a smorgasbord table set for mule deer to enjoy. The mountain lions follow their prey into town, sometimes with disastrous consequences for people, pets, and mountain lions alike.

Birds

There are approximately 250 birds species that spend all or part of each year in the Rocky Mountains. Many of these birds spend the majority of the year at low elevations and only enter the mountains in summer, but some are year-round "mountain birds."

The best way to discuss the birds of the Rockies is according to the places one is likely to see them. For the most part, birds and their habitats are linked. Let's visit the serene Rocky Mountain lakes as we begin to explore the lives of mountain birds.

Lake Birds

One of the best-known lake birds in the Rockies is the common loon *(Gavia immer)*. The behavior of loons on lakes appears sometimes to be quite frantic, and their song is peculiar, ending in what sounds like a bout of maniacal laughter. These features gave rise to the saying "crazy as a loon." The common loon is a tremendous swimmer, able to stay beneath the water for distances of up to 320 feet (100 m).

Loons are residents of montane lakes in the northern Rockies. Loons have a reclusive and solitary nature and prefer to live on secluded lakes in the Rockies. They are highly territorial, and it is rare to see more than one family of loons on a body of water, unless it is a large lake. Loons are excellent swimmers, but they are heavy-bodied birds that are very awkward on land, so they are vulnerable to predation there by coyotes, foxes, and others. Because of this, loons pick nesting sites as close to the water as possible, often on small islands in the lake. They are very graceful swimmers and fliers, but their take-offs and landings are almost comical. Takes-offs involve lots of thrashing in the water, and they require a length of hundreds of feet to get airborne. Their landings are essentially controlled crashes into the water.

Loons lay two eggs per year in a nest made of twigs, grass, and mud. It takes twenty-six to thirty-one days for the eggs to hatch, and another seventy-five to eighty days for the chicks to fledge. In the meantime, the chicks often accompany their parents as they swim by riding on the parent's backs. The loon diet consists mostly of fish, but they also eat aquatic insects.

In spite of their popularity, scientists still have much to learn about the life history of common loons. Even some basic aspects of their lives, such as their age at first breeding, life span, fidelity to mates and breeding territories, and migratory routes remain largely unknown. We do know that loons breed across much of northern North America, from the edges of the arctic tundra to the Rocky Mountains as far south as Yellowstone National Park. During their fall and spring migrations, these loons are often seen on lakes outside the breeding range. They have been spotted as far afield as Nevada and the Mississippi Valley. In the winter, common loons spend their time at sea, near the coasts of California, western Florida, and the Atlantic seaboard.

In addition to their striking black and white plumage, loons are probably best known for their weird and wonderful calls, as mentioned earlier. They produce four major call types: wails, yodels, tremolos, and hoots. Wails are howl-like calls that can travel great distances across the water. These calls keep loons in contact with one another, and once contact is made, several loons get in on the "conversation." Yodels are complex calls made only by males, used to defend their territories. Each male loon can be identified by a unique yodel "signature." Tremolos sound to the human ear like laughter. Tremolo calls are made in response to a threat, but males and females also use them in a duet that may reinforce their pair bond or announce their nuptials to the neighborhood. Hoots are intimate calls that occur between members of a loon family.

The Canada goose *(Branta canadensis)* is both a resident breeder and a migrant in the Rockies. This is probably the best-known North American goose, because it travels far and wide across the continent, honking as it goes. They are found throughout the Rocky Mountain region. Many individuals migrate north in the summer to the arctic coast of Alaska and northern Canada, where they breed and raise their young through the fledgling stage. Increasing numbers remain year-round in low-elevation regions of the Rocky Mountains and elsewhere.

Canada geese are found in a variety of habitats near water in the Rockies, but mainly near standing water, such as lakes and ponds. Females pick nesting sites where they will have good visibility but in isolated locations that provide some protection from danger. Abandoned muskrat lodges are one of their favorite nesting sites. The area also must have access to open water with low banks, so the geese can forage on aquatic plants. Places such as swamps, marshes, meadows, and lakes are among some of the favorite nesting areas. Their nests are made from pond weeds, twigs, grass, moss, needles, and other plant material. Once the eggs are laid, the nest is lined with the mother's feathers and down, insulating the eggs against extremes of temperature.

The Canada goose is mainly a grazer, preferring young, tender plants. They eat many kinds of grasses and forbs, both terrestrial and aquatic. They also eat small numbers of insects, mollusks, and small crustaceans. They feed both on land and in shallow water, where aquatic plants are most abundant.

Females start laying eggs during the first weeks of March and continue as late as June in parts of the Arctic. The egg-laying process is quite protracted, taking a day and a half. The average number of eggs is five, but may be as many as nine. It takes from twenty-five to twenty-eight days for the eggs to hatch. The female sits on the eggs and the male protects the territory from predators.

The male geese are more aggressive than females, using their bills to attack rivals and defend against predators. Canada geese take flight when danger approaches, but may also lay still on the ground when danger approaches. They fly in flocks in the form of a "V" or a diagonally straight line to minimize the effort for the birds following the leader. The birds do not fly right behind the others but off to an angle. They migrate at a slow pace, stopping along the way. Because of this pace, they arrive at the breeding grounds in good physical shape. Most geese mate for life.

Ducks are ubiquitous on Rocky Mountain lakes, ponds, and streams, but not many are true residents of the mountains. One particularly interesting species is Barrow's goldeneye *(Bucephala islandica)*. Most ducks nest near the water, but not Barrow's goldeneye. This duck usually lays its eggs in a hole in a tree. Abandoned woodpecker nest holes work well for Barrow's goldeneye nests. They lay from six to fifteen eggs, which incubate for thirty-two to thirty-four days in the nest. They lay only one clutch of eggs per year. The curious habit of a duck building its nest in a tree serves a useful purpose. It protects the eggs from predators, offering a safe, stable environment for their incubation. Goldeneyes may lay their eggs in a tree that is 1½ to 2 miles (2 to 3 km) from any body of water. Once the eggs hatch, the mother duck coaxes the hatchlings to jump from the nest to the ground, a vertical distance of up to 32 feet (10 m). Fortunately, these little balls of fluffy feathers tend to bounce when they hit the ground. After she rounds up the jumpers, the mother then guides her offspring to the nearest lake. All of this takes place just a few hours after hatching. It takes an additional fifty-five days for the ducklings to fledge (get their adult plumage).

As its name implies, this duck has a bright, gold-colored eye. The males are strikingly colored, with a black head that has a white crescent on the cheek. In the summer, this duck is at home on lakes and rivers in the Rockies. It is most common in the northern Rockies, north of Banff, Alberta. The Rockies populations head west in the winter, spending the cold months in Pacific coastal waters, although some remain in the Rocky Mountain region year-round. This species feeds on dragonfly and damselfly (family Odonata) larvae, which they dive to the bottom to obtain. They also eat pondweed (everybody needs a little bit of green in their diet), mollusks, and the occasional small fish.

Forest and Grassland Birds

The montane forests and adjacent grasslands offer a wide variety of habitats and food resources for birds. The conifers provide not only nesting sites but also food for many species, either directly or indirectly. Pine cones are the source of nutritious seeds, and the insects that attack conifers are the prey of several species of woodpeckers (family Picidae). Montane forest rivers and lakes provide fish for birds of prey, including the bald eagle *(Haliaeetus leucocephalus)* and the osprey *(Pandion haliaetus).*

The bald eagle once fished the waters throughout North America. It was nearly extirpated from the lower forty-eight states by the beginning of the 1970s because of the damaging effects of such pesticides as DDT, dieldrin, and endrin. These poisons made their way from agricultural fields into streams and lakes, where they were taken up by fish. When eagles ate the fish, they began accumulating the pesticides in their tissues. The decline in eagle populations was a result of the effects of the pesticides on eggshell development in females. Females exposed to the pesticides were laying eggs with shells that were too thin and fragile, so chicks did not survive. By 1963, the number of breeding pairs of bald eagles in the lower forty-eight states had declined to fewer than 500. Bald eagles came under federal protection in 1967, six years before the enactment of the Endangered Species Act. DDT was banned in the United States in 1972; dieldrin and endrin followed shortly thereafter. By 1974, 700 pairs of bald eagles were counted; almost half of these birds were living in the Great Lakes region. Since these pesticides were banned by the U.S. government, eagle numbers have begun to rebound. A 1994 census found about 4,500 pairs in the lower forty-eight states, and these eagles are recovering well in the Rockies. The species itself was never really threatened with extinction, however, because large, healthy populations of bald eagles have persisted in regions of Canada and Alaska where agriculture is impractical.

The golden eagle *(Aquila chrysaetos)* is another magnificent bird of prey that is common in the Rockies. It is nearly as large as the bald eagle, with a wingspan of up to 6½ feet (2 m). The bird gets its name from the mantle of golden feathers that clothe the back of its head and neck. The legs of the golden eagle are covered in feathers; this is one feature that separates it from the bare-legged bald eagle. This bird nests high up on rocky ledges, and soars for hours over the mountain terrain, looking for prey. Golden eagles typically fly at about 30 miles (50 km) per hour, but they can attain speeds of up to 125 miles (200 km) per hour when chased or when diving after prey. They feed mainly on rodents, rabbits, and hares, but have been seen trying to take young mountain goats and mountain sheep.

Golden eagles nest in the spring, and the males and females pair for life. They build enormous nests made of sticks. These nests may be 10 feet (3 m) in diameter and 3 feet (1 m) thick. The female lays two eggs, either once a year or once every two years. The eggs hatch in forty-three to forty-five days, but one egg hatches one or two days earlier than the other. The older chick usually kills the younger one. It takes sixty-six to seventy-five days for golden eagle chicks to fledge.

Rocky Mountain populations of golden eagles overwinter on the Pacific Coast of Canada and in many regions of the United States. Most golden eagles from the central and southern Rockies do not move far in winter. Recently, wildlife biologists discovered that golden eagles migrate in groups as they return to the Canadian Rockies in the spring. From early March to early May, 6,000 to 8,000 eagles move north through the front ranges of the Rockies, flying in groups of ten to forty individuals. Sometimes, other birds of prey join in these groups, such as bald eagles, hawks, and falcons. As many as 1,000 golden eagles have been seen arriving in the Canadian Rockies in a single day.

The American kestrel *(Falco sparverius),* a much less formidable bird than either of the eagle species, is a small, common falcon in the Rockies. In fact, it is one of the most common birds of prey in the area. American kestrels are about the size of the American robin. They have a wingspan of 21 inches (53 cm), making them the second smallest kestrel in the world. This bird is sometimes called a "sparrow hawk." This name is misleading, because this bird neither preys on sparrows nor is a true hawk. Its habitat includes mountain meadows and grasslands adjacent to the Rockies. Its call is a high-pitched "killy-killy-killy." This bird nests throughout the Rockies and winters as far north as the Central Rockies. Kestrels are partially migratory. Northern birds migrate south, as much as 2,000 miles (3,200 km). Southern ones may migrate a short distance or not at all.

Kestrels eat mostly insects but occasionally take small animals, including rats, mice, birds, and reptiles. There is a seasonal pattern to their diet. In summer they feed mainly on insects that they catch either on the ground or in the air. Wintering birds feed primarily on rodents and birds. The kestrel's hunting method is quite distinctive, which makes them easy to spot in the wild. They often hover, almost motionless in the sky, as they search for prey. Hovering allows a kestrel to maintain a stationary position aloft, from which it can scan the ground below. However, when there is no wind, hovering is too energy-expensive, so the kestrel may perch in trees or other elevated objects to search for prey.

The male kestrel puts on courtship aerial displays of steep climbs and dives, and it brings the female gifts of freshly killed prey. Pair bonding among kestrels is usually permanent. A pair is established after the male takes over a particular territory. Afterward, a female will begin to hunt and associate herself with the male. The major components that strengthen the pair bond include courtship feeding of the female by the male, aerial displays, and the search for a nest site. An existing tree cavity, often a woodpecker hole, is chosen for nesting in late April or early May. Four to five eggs are laid in an unlined nest. While brooding the eggs, the female is fed by the male, who brings gifts of freshly killed prey. Once the eggs hatch, the parents share the job of feeding the young. The chicks hatch in about twenty-eight days, and fledge about thirty days after hatching. They grow very quickly, reaching adult weight in about seventeen or eighteen days.

One of the most magnificent hunters of the montane forests is the great horned owl *(Bubo virginianus).* With a wingspan of up 5 feet (150 cm), the great horned owl is the largest owl in the Rocky Mountains. The conspicuous ear tufts or "horns" give the species its name. This owl is widespread from Alaska and northern Canada south throughout the Americas. In the southern Rockies the great horned owl lives at elevations up to 10,000 feet (3,000 m). They are year-round residents and very long-lived. Great horned owls in captivity have been known to live as long as sixty-eight years. They live in both forests and grasslands, and nest in trees, caves, cliff ledges, or even on the ground. They will often use an abandoned nest of a large bird such as hawk or a crow, but their preferred nesting site is a large, hollow tree. The female owl is always bigger than the male. These are very powerful raptors. The strength of their grip may equal that of the bald eagle. Wherever horned owls live, they are the dominant bird of prey in that area.

This owl is nocturnal. Its dark feathers keep it well-camouflaged at night, and its silent flight enables it to surprise its prey. Great horned owls are not fussy about their prey. They will attack almost any living prey, including frogs, mice, voles, rats, rabbits, birds (including other owls), reptiles, and fish. They are among the few predators of skunks, whose scent often remains on the owl's plumage. Indigestible portions of their food, such as bones, hair, and feathers, are compressed and regurgitated as compact pellets.

The great horned owl is among the earliest birds to nest in North America. The incubating parent is often snow-covered as it sits on the nest. In low-elevation regions, they begin to breed as early as January. In the Rocky Mountains, they begin in March, which is the snowiest month in many parts of the Rockies. Mountain-dwelling great horned owls lay more eggs than their lowland counterparts, up to fifteen eggs in a clutch. However, only two to five of these eggs hatch. The eggs hatch in about twenty-eight days, and the young owlets get their adult feathers thirty-five days after they are born. The young may be cared for by the adults for five months. Because it is so cold at the time of nesting, incubation begins immediately after each egg is laid. As a consequence, the eggs hatch in sequence, which gives the first hatchling a size advantage over its siblings. If food is scarce, the largest will outcompete the others for food and only that one survives.

These owls do more hooting during their mating season than any other time of the year. So if you visit the Rockies during the springtime and would like to find out if there are any great horned owls in your vicinity, just go out after dark, cup your hands around your mouth, and let out a loud hoot. Wait a few seconds, listen carefully, and try again. If there are any great horned owls in the area, they will very likely hoot back at you or even fly over to find out who is invading their territory.

In contrast to the powerful great horned owl, hummingbirds (family Trochilidae) are a delicate delight to summer visitors in the Rockies, so much so that many mountain cabins have hummingbird feeders hanging from their porches or eaves. These tiny, beautiful creatures seem almost too delicate to survive the cold mountain nights.

There are four species of hummingbirds commonly seen in the Rockies. All are mostly clothed in brilliant green plumage, and none spends much time at elevations higher than the lower montane forest. As you sit on a porch, you can try to distinguished one from another as follows: The black-chinned hummingbird *(Archilochus alexandri)* has, as the name implies, a truly black throat; the broad-tailed hummingbird *(Selasphorus platycerus)* makes a metallic whistling sound with its wings as it flies; the rufous hummingbird *(Selasphorus rufus)* has a solid red back; the calliope hummingbird *(Stellula calliope)* is smaller than the others and the males have colored throat feathers that form streaks against a white background. All four are only summer residents in the Rockies. Hummingbirds feed on nectar, reaching deep into tubular flowers to lap up the sweet juices with their long, slender bills. Although they do not fly very high or very far in a day, their wings beat so fast and they are such agile fliers that they can easily hover over a flower and can even fly backwards.

Despite their diminutiveness, which tends to make them look harmless, hummingbirds are highly territorial birds that are quite aggressive toward each other. They spend a great deal of their time in aerial combat, either defending their own territory or trying to horn in on some other hummingbird's turf. They lunge and feint like fencers, using their long, pointed beaks as swords.

There really is a yellow-bellied sap sucker *(Sphyrapicus varius)* and it is common in the coniferous forests of the Rockies. This member of the woodpecker family uses its sharp, stout bill to drill holes in trees that ooze the sap it uses for food. It prefers the sap of aspen, birch, and poplar trees. The holes are drilled in horizontal lines around the tree, and they angle down into the bark, creating tiny reservoirs in which the sap pools. This bird has a special brush-like tongue that it uses to lap up the sap (it might therefore be better called a sap-lapper than a sap-sucker, but the latter name has a nice alliterative sound). The sapsucker's summer range extends throughout the Rockies, but the northern tip of its winter range is in New Mexico.

Another common representative of the woodpecker family likely to be seen in the Rockies is the downy woodpecker *(Picoides pubescens)*. Downies differ in appearance from sap suckers. Downies have a white patch in the middle of the back, whereas sapsuckers have a white patch on each wing. Downy woodpeckers are the smallest woodpecker in the Rockies and are year-round residents. They climb trees, probing for insects and their larvae hidden in or under the bark. Their tree-climbing ability is aided by anatomy in a couple of ways. First, they have four-toed feet with two toes facing forward and two facing backward. This helps the woodpecker to hold on tight to the vertical surface of the tree. They also use their stout tail to brace themselves as they shift their feet from one place to the next as they climb. These climbing adaptations, along with a sharp, stout beak for probing and drilling into bark, make this diminutive woodpecker an effective insect predator in the montane forests of the Rockies.

The common raven *(Corvus corax)* is, to my way of thinking, an unusual bird. The biggest of the "songbirds," it haunts the woods of the Rockies and the boreal forest with its loud, raucous calls. These birds follow you around as you walk through the woods, squawking in a way that seems to many Native peoples to be a kind of species-to-species communication. In fact, ornithologists have catalogued more than thirty distinct calls made by this species. No wonder the raven is considered by many Native peoples to be a messenger from the spirit world, a wise, crafty bird to be respected.

Ravens are naturally curious. They explore their territory, keeping a sharp eye out for anything new in the environment. This general principle has been known for a long time, but lately scientists have been learning just how curious these birds can be. In one study, four young ravens were observed over a period of several weeks.

During that time, the birds examined 980 different objects in their territory, including ninety-five different kinds of objects. This contact ranged from poking at objects to trying to taste them. The scientists then introduced forty-four new items into the ravens' environment. The birds checked out the new items as soon as they appeared, but items they deemed inedible were soon ignored. On the other hand, edible items became instant favorites. It is easy to see how this inquisitive behavior results in ravens finding and eating nearly all edible objects in their territory.

Ravens are quick to find and exploit dead animal carcasses (carrion) on the landscape. A crowd forms quickly, because the first raven on the scene produces a series of very loud cries, called "yells," that let other ravens know of the carrion. These yells mostly attract the attention of wandering ravens, rather than family members, so most ravens at carcasses are not in the company of their close kin. This behavior is unlike that of some other bird species that act first to benefit their own relatives. So even though ravens appear to us to have a self-centered attitude (probably because of their aggressive vocalizations and lack of apparent fear in the company of humans), they can also act unselfishly when there is carrion available for a communal feast.

Another noisy denizen of the montane forest is Steller's jay. All members of the crow family seem to be endowed with a certain attitude about life. They are often aggressive, loud, and unafraid of humans. Steller's jay is no exception. Unlike the black, somber coloration of the raven, this jay has a black head and neck contrasting with a brilliant blue lower torso and wings. Steller's jay is a medium-sized member of the crow family, with a wingspan of 18 to 19 inches (45 to 48 cm). It lives in montane and subalpine forests throughout the Rockies, and it ranges north into the boreal regions of Canada. The Steller's jay usually lives in open woodlands, at the edges of clearings, and along forest waterways. It is a year-round resident in the Rockies, but it migrates upslope in the summer and down to lower elevations in winter. Populations in the northern Rockies of Canada have been tracked moving south by as much as 105 miles (170 km) in winter.

Steller's jays nest in conifers, close to the trunk, near the top of the tree. They build their nests of conifer branches and twigs, and line them with dried grasses, mud, and dried mosses. They mate for life, and both individuals participate in selecting a site and building the nest. The eggs are laid between April and early July. Most clutches have four eggs that hatch in about sixteen days. The young remain in the nest for about twenty days.

The Steller's jay is a highly social bird. Flocks of various sizes form often, with mates rarely parting. Steller's jays join forces to harass birds of prey. They "mob" these predators by flying at them while making loud calls. The Steller's jay also mimics the call of the red-tailed hawk to scare off other species. Its own call is a sharp "hack, hack hack!"

Four species of chickadee *(Parus)* are found in the montane forests of the Rockies. Three of the four give distinctive calls of "chick-a-dee-dee-dee." But in addition to this familiar call, chickadees also have one of the largest vocabularies of other calls. More than fifteen different chickadee calls have been identified by ornithologists. These birds hop busily from branch to branch of conifers, in search of food from dawn to dusk. The chickadee feeds on insects, seeds, and berries. They prepare their nest with grass, fur, plant down, feathers, and moss in a hole in rotten tree stumps, natural tree cavities, or abandoned woodpecker holes. This bird is constantly active, either hopping, clinging, or hanging from tree branches.

Chickadees are usually seen in pairs or small groups. When nesting is over and the young are fledged, chickadees form small flocks of eight to twelve birds that roost and forage together until spring. Finding food in the winter is often difficult, so group foraging increases their chances for success. Flocking together also helps chickadees keep a look-out for predators. Wintertime in the Rockies can be very quiet. A walk through the woods at this time of the year will not usually offer much for the bird watcher, as most of the birds have abandoned these chilly, snow-covered mountains for warmer climes. But you can almost count on hearing the bright, cheerful call of the chickadee. They're out there, even in the coldest weather, small in size but indomitable in spirit.

Unlike the chickadees, the warblers (family Parulidae) of the Rockies migrate south or to the lowlands for the winter, so they are not year-round residents. The warblers of the Rockies are small, brightly colored songbirds with either all or part of their bodies clothed in yellow feathers (contrasting with the black-and-white coloration of the chickadees). As the name implies, warblers often have a warble, or trill, in their songs. For instance, the yellow-rumped warble has a soft, warbling call that sounds to human ears like "sweet-sweet-sweet-peachy-peachy." Wilson's warbler *(Wilsonia pusilla),* another common Rocky Mountain species, has more of an edge to its song, which has been described as "CHI-CHI-CHI-CHI-chet-chet." Warblers feed more on insects than on seeds and berries. Some are quite adept at snatching flying insects out of the air, while others pick them off of the trunks of trees and some specialize in picking caterpillars off of leaves. These birds are little and shy. They are not often seen, as they hide in thickets and treetops. The easiest way to find them is by their distinctive calls.

Tundra Birds

The alpine tundra, with its frigid, windswept landscapes, might seem an inhospitable place for birds, yet several species spend the summer here, and a few hardy types even spend the winter on the tundra. A species of grouse (family Tetraonidae) that inhabits the alpine tundra of the Rockies demonstrates how birds can successfully live in this harsh environment. The willow ptarmigan *(Lagopus lagopus)* is essentially an arctic grouse species that has a southerly range extension into the northern Rockies of Canada. Its principal food is willow leaves. This is a remarkable bird, able to withstand the coldest climates of the continent. They are feathered from head to toe, and they burrow in snow to get out of the icy blasts of winter wind that would otherwise kill them. They have mottled brown or reddish-brown plumage in summer, providing good camouflage in the rocky, brown, tundra landscape. However, in winter the willow ptarmigans molt their summer plumage and both the replacement feathers and those that are retained are white, blending in with the snow. When approached, these birds stand perfectly still, hoping to avoid detection. They are poor fliers and only attempt to fly away from trouble at the last second. Willow ptarmigans are year-round residents of the tundra.

Some finches (family Fringillidae) are also tundra dwellers. In summer these birds forage for insects, berries, and seeds on the alpine tundra. Their long, pointed wings help them to maintain flight control in the perpetually windy tundra landscape. These birds do not fly high; in fact, they stay pretty close to the ground. The reason is simple: If they get more than a few feet off the ground, the almost perpetually high winds will blow them off the mountaintop, down into the forests, and hence out of their habitat.

Another set of adaptations for life on the tundra concerns the reproductive cycle of these finches. The summer season is extremely short on the tundra, so every part of the cycle must be accomplished quickly. For instance, the brown-capped rosy finch *(Leucosticte australis)* incubates its eggs for only twelve to fourteen days, and once the eggs hatch, the young are fully fledged in as little as sixteen days. Thus the whole cycle from egg-laying to maturation of the young to the point they can fly takes as little as twenty-eight days. The chicks are given lots of food in the nest, which helps speed their growth and development. Their parents have cheek pockets that allow them to bring large quantities of food to the chicks. So when the cold weather comes, both generations are ready to fly down to lower elevations, where they spend the winter. It seems that the ptarmigan are the only birds adapted to alpine winters in the Rockies.

"Cold-Blooded" Animals

The animals described in this part of the chapter are not able to regulate their internal temperatures by their own physiology. They must modify their behavior to avoid extremes of heat and cold, which means they seek shady shelter to avoid the hot summer sun and bask in the sun on cool mornings so that their bodies will absorb enough heat to allow them to get moving. This way of life is widely, if inaccurately, termed "cold-blooded." Of course "cold blood" is one thing these creatures try to avoid (along with overheated blood). We will now look at the life histories of a few species of fish, amphibians, reptiles, and insects. I think you will be amazed at some of the ways these animals have of operating in the perennially cold climate of the Rocky Mountains.

Reptiles and Amphibians

The reptiles and amphibians of the Rockies are few but fascinating. Although they are unable to control their internal temperatures except through external means, some species have found ways of coping with life in some very cold climates, such as the subalpine zone of the Rockies. The skin of all amphibians is vulnerable to dehydration, but the species that live in dry climates in the Rockies manage to keep themselves moist by living in wet patches and by avoiding exposure to the sun. In other words, they have created a moist habitat for themselves where there is little moisture available, strictly through their behavior. Along with their natural roles in ecosystems, the amphibians now play an additional role: that of pollution watchdogs. These moisture-loving creatures are very sensitive to both air and water pollution, and the regional disappearance of some species is a harbinger of environmental hazards that must not be overlooked.

There are seventeen native amphibians and twenty-nine native reptile species in the entire Rocky Mountain region. Most live on the periphery of the mountains, in adjacent plains and foothills. Nine amphibians and six reptiles are true mountain dwellers. Let's focus on these mountain-living reptiles and amphibians.

Amphibians are closely tied to water. Most lay their eggs there and for many species the immature stages develop there. Many adult amphibians inhabit aquatic or moist places, such as damp forests, swamps, bogs, ponds, and lakes. The Rockies amphibian fauna includes salamanders, toads, and frogs.

There are only a few salamanders that live in the Rockies. These need very moist habitats, and few parts of the Rockies offer the kind of consistent, damp conditions they require. It is little wonder that evolutionary biologists imagine something akin to a salamander as the first kind of creature to make its way out of the water and establish a foothold on the land. Salamanders appear to be essentially water creatures that find ways to get up onto the land, find something to eat, and find their way back to the water. This is an oversimplification of their lifestyle, but salamanders do go to great lengths to keep themselves moist. One interesting and colorful species is the Pacific giant salamander *(Dicamptodon ensatus).* Although its name suggests otherwise, Pacific giants grow only to about 6 inches (15 cm) in length. This is larger than average for salamanders, but somehow the name "giant" seems a bit of an exaggeration. This is a species of the Pacific Northwest region that ranges east into the Rockies of Idaho, where it can find the type of moist habitat it requires. Unlike most salamanders that come out from cover only at night, this species wanders about both day and night. It seeks cold stream headwaters in which to lay its eggs in the spring.

The tiger salamander *(Ambystoma tigrinum)* might be more appropriately called a giant. This salamander occurs throughout the Rocky Mountain region, and reaches lengths of 7 to 12 inches (17 to 33 cm). It ranges from the plains up to 11,000 feet (3,350 m) in the Rockies. It needs moisture in its habitat, but is able to survive in relatively dry localities by its behavior. In other words, it has found ways of "cheating the system." It lives in some dry regions but does not expose itself to that dryness. During the day, it seeks shelter in damp places, such as under logs near ponds or streams and in burrows. It comes out only at night, and generally stays near water. The only time it moves great distances is when it migrates into new regions, and it does this only immediately after a heavy rainfall, when the ground is particularly wet. It lays its eggs in still waters. In these ways, the tiger salamander stays moist and cool, even in relatively hot, dry climates.

The tiger salamander's food source consists mainly of worms, snails, insects, and slugs, but they are also known to take smaller salamanders, frogs, mice, and small fish. Adults live underground for most of the year and usually dig their own burrows, unlike other species that use burrows of other animals. They have been found more than 24 inches (60 cm) below the surface.

The toads and frogs are more often seen than the salamanders. These hopping amphibians are an important part of many ecosystems, especially lakes, ponds, and marshes, where they sometimes reach impressive numbers. The two species of toads that are native to the Rockies are the boreal *(Bufo boreas)* or western toad and Woodhouse's toad *(Bufo woodhousei).* The boreal toad occurs throughout the Rockies, except New Mexico. As might be expected from such a wide-ranging species, it inhabits a variety of habitats, including streams, springs, grasslands, woodlands, and meadows. It generally stays near water. In regions of warm climate it is active only at night, but in the cooler climes it is also active during the day. Unlike many toads and frogs, the boreal toad tends to walk rather than hop. Unfortunately, this toad is becoming scarce in some regions. In 1996, the Colorado Division of Wildlife began hanging "Wanted" posters for this toad, in an effort to familiarize the general public with its appearance, so that they would report seeing it. The poster says, "If you see this toad, its eggs, or its black tadpoles above 8,000 feet elevation, please contact (the Division of Wildlife)." The boreal toad is an endangered species in Colorado. The once common species is now becoming quite rare, and the reasons for its decline are unknown.

Any toad found above 8,000 feet (2,440 m) is probably a boreal toad. It is the only toad indigenous to the high elevations of the Rocky Mountains. The tadpoles are more easily spotted than the adults, with their jet black skin and their tendency to linger in warm ponds in less than an inch of water. The toad tadpoles are easily distinguished from other tadpoles, because boreal tadpoles have eyes that do not protrude from the margin of their heads.

Once known to occur in twenty-five of sixty-three counties, and potentially in seven others, the boreal toad is absent in more than 83 percent of previously known locations in Colorado. There are now fifty breeding sites left in Colorado and a single breeding location in Wyoming, but most of the sites have only a few breeding adults. Several efforts to reintroduce the toads have taken place over the past few years, but all have failed. Though the reason for their decline remains somewhat of a mystery, researchers have discovered that a species of fungus *(Batrachochytrium dendrobatidis)* has been attacking boreal toads in the Rockies, possibly as far back as the 1970s. These fungal attacks appear to be fatal to the toads.

The leopard frog *(Rana pipiens)* is found throughout the Rockies, except in central Idaho. Indeed, it is the most wide-ranging amphibian in North America, found in deserts, mountains, prairies, and coniferous and deciduous forests. Leopard frogs vary considerably in length, from 2 to 5 inches (50 to 130 mm). They live in and around almost any freshwater habitat, from streams and rivers to ponds, lakes, and bogs. Another name for the leopard frog is "meadow frog," and during the summer these frogs may wander far away from water. In Colorado this species can be found at elevations up to 11,000 feet (3,350 m). In Alberta, most live in clear, clean freshwater springs in lightly wooded areas. They are most active after sunset in warm, wet weather. They winter under stones in the moving water, as long as there is enough oxygen.

The tadpoles eat plants, algae, dead tadpoles, or small invertebrates, such as insects. Adults eat almost anything they can catch, including insects, small vertebrates such as mice or fish, and even other frogs.

Leopard frogs do not breed until they are three to four years old. In the spring, males begin calling to attract females. Breeding begins in May at low elevations in the Rockies and as late as June at high elevations. Each female deposits up to 3,000 eggs in a large, flattened mass of dark jelly in the water. The eggs hatch into tadpoles in ten to twenty days. By early August, they transform into adults.

As ubiquitous as this frog is, it too is being threatened by pollution in the Rockies. The Montana Natural Heritage Program has documented the loss of this species from eight out of nine historic sites in the state.

Reptiles

Unlike amphibians, most reptiles such as snakes and lizards are less dependent on water. Of the more than fifty species of lizards that inhabit western North America, only seven are true mountain inhabitants of the Rockies. One of these, a variety of the eastern fence lizard, is found in New Mexico, Colorado, and Wyoming. You can see this lizard in a wide variety of habitats, including forests, woodlands, and prairies. This lizard is found in sunny, rocky terrain. In Colorado, it is found up to 9,200 feet (2,800 m), although it is more commonly found below 7,000 feet (2,130 m). The adult males are easy to spot. In the summer (mating season) they sport metallic blue markings on their head and neck regions, presumably to attract females. These fence lizards eat a wide variety of insects, including ants, beetles, grasshoppers, and butterflies and moths.

There are ten snake species that live as true mountain dwellers in the Rockies. One of the more surprising snakes that resides in the Rockies is a boa. Many people think of tropical rainforests when they hear about boas, pythons, and other constrictors (snakes that kill their prey by squeezing them to death), but there is a native boa in the Rockies called the rubber boa *(Charina bottae).* It does not live in rubber trees but rather its smooth texture and flexibility give it the look and feel of rubber. The rubber boa is a northern snake, found in central and northern parts of the Canadian Rockies. It inhabits the banks of rocky streams in both grasslands and woodlands, up to an elevation of 9,000 feet (2,750 m). It feeds on small mammals and lizards.

The racer *(Coluber constrictor)* is a member of the family Colubridae. It is a slender, fast-moving snake that can grow up to 78 inches (198 cm) long. Racers are active during the day. They live in open-ground habitats in both prairies and forests but do not live in high mountains. The species is fairly wide-ranging in the Rockies, from New Mexico to southern Idaho and Montana. It is most often found on the ground, but it also climbs trees and shrubs. It hunts in sunlit meadows where lizards come to bask, and also eats frogs, small mammals, and insects, as well as birds, reptiles, and amphibians. Although a member of a family of constrictor snakes, the racer does not use constriction to kill its prey, but it may pin its prey down with its body. As the name implies, these snakes are fast movers. A subspecies from the eastern United States, the black racer has been clocked moving more than 30 miles (50 km) per hour.

There are three species of venomous snakes that inhabit the Rockies, all of which are in the viper family: the massasauga *(Sistrurus catenatus),* the western diamondback rattlesnake *(Crotalus atrox),* and the western rattlesnake *(Crotalus viridis).* Western rattlesnakes roam the Rockies from New Mexico to southern Alberta and range up to 12,000 feet (3,650 m), making them the perfect species to illustrate rattlesnake lifestyles in the region. This snake frequents rocky outcrops and often dens in rock crevices and shelters, especially toward the northern end of its range. The western rattlesnake is generally active from April through September, when the air temperature is above 50°F (10°C). Rattlesnakes conserve their energy during the coldest winter days and nights by entering a state of inactivity, called torpor. When the warm weather arrives in spring, they migrate long distances to and from their dens. This snake often prowls during the middle of the day. When summer temperature get too hot (above about 77°F or 25°C), the snakes avoid the heat of mid-day by seeking out shady places to rest.

Pit vipers (family Viperidae), like the western rattlesnake, use a couple of hunting strategies: actively searching for prey animals and sitting and waiting in ambush. The former method is used when prey animals are scattered; the latter is used when the snake manages to find a ready source of food, such as a colony of rodents. Rattlers detect their prey by vision, by infrared (heat) detection, and by scent. They really do not need to see their prey, because even a blind rattlesnake can accurately direct its strikes on a prey animal. Odors are detected through the snake's nose and on the tongue, which is flicked out to "taste" the scents on the air, then brought back in, where it rubs over a specialized scent-detecting organ on the roof of the mouth. The western rattlesnake uses a pit organ located in a patch of scales on its head to detect the body heat of its prey. The heat-sensing ability is limited to a range of about 14 inches (35 cm), but it helps the snake focus on prey it has already detected through vision and scent. The snakes can detect minute differences in temperature, allowing them to differentiate between a deer mouse and a warm rock nearby.

Once the prey is found, the western rattlesnake uses efficient and deadly venom to subdue it. The snake's long fangs are individually hinged so that they can easily rotate into biting position. At the base of each hollow fang is a duct leading to the venom gland. The venom serves two purposes: to quickly subdue and kill the prey and to aid in digestion. Surprisingly, the western rattlesnake is not immune to its own venom, unless it swallows it along with a meal, so that it can be broken down in the stomach. But if that venom is injected into the snake's bloodstream (as it would be if bitten by another rattler), it can be fatal. Western rattlesnake venom is more toxic than that of its larger relative, the western diamondback rattler. This venom potency, combined with an irritable disposition, makes the western rattlesnake a very dangerous creature.

The western rattlesnake has the most variable diet of any rattlesnake in North America. They seem to prefer warm-blooded prey, but they will also take reptiles, amphibians, and bird eggs. Prey selection is limited by the size that can easily be swallowed, and prey size increases as the snakes grow. Rodents are the mainstay of this species' diet, including ground squirrels, kangaroo rats, young cottontail rabbits, mice, pocket gophers, and voles.

Contrary to popular opinion, you cannot tell a rattlesnake's age by the number of rings on its rattle. Rattlesnakes add another ring to the rattle on their tails each time they shed their skin, which is generally three times per year. The rattle is composed of hardened keratin, the same material found in human fingernails. As the snakes get old, the outermost rings tend to break off, muddling the picture even further.

Our three venomous species are not as deadly to humans as some other venomous snakes, but they must be treated with respect nonetheless. When you see or hear one (the rattling sound is often the first indication that you are close to a rattlesnake), stand still until you have spotted it; do not jump or run blindly or you may run right into it. When the snake has been located, back away from it, avoiding sudden, jerky movements that might cause it to strike. Also, do not handle dead rattlers. Some people have been bitten when the reflex action of the jaw muscles causes the jaw to contract.

Insects

In one of the back rooms of the Smithsonian Institution's National Museum of Natural History in Washington, D.C., a large banner once proclaimed in tall, red letters:

INSECTS RULE!

As someone trained in entomology, I find it hard to disagree with the banner, yet so many simply lump everything from beetles to bumblebees into a category called "bugs" that entomologists feel the need to tell the world about the fascinating diversity of these animals. We study these tiny animals because we find their anatomy, physiology, and ecology fascinating. Insects play vital roles in virtually all terrestrial ecosystems. There are so many kinds of insects that we have yet to discover them all, even in such well-traveled regions as the Rockies.

No one has completed a faunal survey of the insect species of the Rocky Mountains (or scarcely any other region, for that matter). But more than 150 years of research of the insect life of the Rockies has given us a wealth of information.

Insects have received the most study of any invertebrate group because of their economic importance to farming and forestry. Among the forty-two orders of insects, we have learned the most about such groups as ants, bees, wasps, grasshoppers, locusts, butterflies, moths, and beetles. Url Lanham, former curator of insects at the University of Colorado Museum, estimated that 10,000 species of insects inhabit

COREL CORPORATION

Ground beetles. These predatory insects live at all elevations in the Rockies.

the Colorado Front Range region. If this estimate is accurate, then at least 20,000 species of insects must live in the entire Rocky Mountain region. Little wonder, then, that science has had a difficult time identifying and categorizing all of these creatures.

The beetles are the largest order of insects. In fact, they are the largest group of organisms on the planet. About half a million species of beetles are known to science, and untold numbers, perhaps millions more, are yet to be identified (chiefly in tropical forest regions). More than 30,000 species of beetles live in North America. This number exceeds the sum total of all flowering plants and all vertebrate animals combined. One family of beetles, the ground beetles (family Carabidae) are an important group of insects throughout the Rocky Mountain region. About 750 species of ground beetles inhabit the Rockies. Ground beetles are mostly generalist predators—that is, they prey on any animals they can catch. A few are specialists, eating only certain prey animals. For instance, beetles called *Calosoma* are caterpillar hunters, and beetles called *Cychrus* have flattened heads and mouth parts that allow them to prey on snails, which they prefer to all other food items. The caterpillar hunters climb into trees, seize a caterpillar, make a neat incision near the head end, peel the caterpillar's skin back like peeling a rubber glove off of a hand, then plunge their mouth parts into the wriggling mass of goo left behind, lapping it up before looking for another victim. The snail hunters face a more resistant prey, because snails retreat into hard shells when disturbed. But *Cychrus* has worked out a clever strategy to overcome this. The beetle waits until the snail retreats into its shell, then it simply steps up to the shell and starts spitting its digestive juices into the opening. Within a few minutes, the snail in the shell has been killed and predigested by these juices. The beetle then laps the liquid *escargot*, using the snail shell as a soup bowl.

Ground beetles vary considerably in size, and they inhabit a wide variety of habitats in the Rockies. The mountain topography creates many different ground beetle habitats, from stream and lake shores to moist meadows, rocky slopes, and shaded forest floors. The substantial diversity of ground beetles in the Rockies is a result in large part to this variety of habitats.

The richest region for ground beetle species diversity is British Columbia, followed by Colorado, Alberta, New Mexico, and Wyoming. This makes good ecological sense when you take into consideration the diversity of habitats available to insects in the various regions. British Columbia is rich in beetle species because it contains a wide variety of biological communities, including moist, cool Columbian forests; the warm, dry Purcell Trench; the boreal forest in the north; and extensive subalpine and alpine regions in the east.

British Columbia and New Mexico have relatively large numbers of endemic (not found anywhere else) species. The reason for this is almost certainly because those two regions tap into the faunas of other ecosystems outside the Rockies proper. British Columbia is home to the rich fauna and flora associated with the moist, moderate climate of the Pacific Coastal region. New Mexico has not only the southern tip of the Rockies but also has considerable desert regions. Colorado, Wyoming, Montana, and Alberta have fewer ground beetle species that are unique to their own region simply because they share most of their fauna up and down the Rocky Mountain corridor and the plains on their eastern slopes.

There are some interesting trends in the number of beetle species of the Rockies.

The southern Rockies have 245 ground beetle species that are not found in the central and northern Rockies. The central Rockies have only five species not found in the northern and southern Rockies, and the northern Rockies have 236 ground beetle species not found in the central and southern Rockies. About one third, or 262 ground beetle species, range throughout all parts of the Rocky Mountains (north, central, and south). Thus if you move north to south along the Rockies, some beetles will accompany you all along the journey, and others will either leave you or greet you as you pass through their region. For my sake, try not to step on them.

Butterflies are another group of insects that are reasonably well-known in the Rockies. It is no coincidence that groups such as beetles and butterflies are better studied than such groups as flies or springtails, as so many amateur collectors have devoted their spare time to collecting these insects in the Rockies. Very few professional entomologists devote their time to collecting specimens, so the collections of dedicated amateurs are vitally important sources of information. Butterflies are probably the most widely collected insect group in the world, because their appearance appeals to many people, and their dried exoskeletons, complete with wings covered in colorful patches of scales, make attractive displays.

Surprisingly, the butterfly fauna of Colorado is more diverse than that of any other state in the union. Martin Brown cataloged 248 species of butterflies in Colorado. The closest competitor was California, with 208 species. Most states have 150 species or fewer. The Canadian Rockies have fewer butterflies than other parts, but Ben Gadd recorded ninety species for Canadian parts of the Rockies. Brown attributed the great diversity of the Colorado butterfly fauna to the four great rivers that have their headwaters there (the Colorado, Rio Grande, Arkansas, and Platte). These rivers tap the lowland faunas of these drainages, and the river valleys serve as flyways by which butterflies move into Colorado.

The butterfly fauna of Colorado is divided into year-round residents and summer visitors. The summer visitors migrate into Colorado to feed on nectar stored in flowers and perhaps to find a mate, but the females return to other regions to lay their eggs. Some butterflies spend the winter in Mexico or along the coast of the Gulf of Mexico. One species that does not quite belong but is found from time to time is the pipe-vine swallowtail *(Battus philenor)*. As the name implies, the larvae of this species feeds on the pipe-vine plant, which is not native to Colorado. However, pipe vine was introduced into the region by horticulturists in the early part of this century, and the butterfly has followed, now becoming well-established in some regions.

Many species of butterflies in the Rockies take advantage of dark-color heat absorption on and near their thoraces and abdomens. Many species have black heads, thoraces, and abdomens, and many have black regions at the bases of their wings. Because butterflies, like all other insects, obtain their body temperature from the environment, they must find ways to warm their bodies using something other than metabolic heat. This is especially true of flying insects that live in cold regions. Their flight muscles must reach a certain temperature before they can fly.

Few butterflies are able to cope with the always-frigid, often extremely windy environment of the alpine tundra. Those few species that live there often have additional dark patches on the undersides of their wings, which they hold over their bodies like a tent as they sit on the ground, allowing the sun to warm their flight muscles. They fly in short bursts, stay close to the ground, and land immediately if the sun goes behind a cloud. Thus they try to keep out of the wind and make the most of the solar heating available to them. This is truly life on the edge.

Pollinators

One of the most important aspects of many species of insects is their role as plant pollinators. Chief among the pollinators of the Rocky Mountain flora are bees, wasps, flies, butterflies, moths, and beetles. The native bees and wasps of the Rockies include solitary bees and wasps, as well as those that form large colonies. The honey bee *(Apis melifera)* is not native to North America but was introduced from Europe in the seventeenth century. Before the honey bee arrived, many other kinds of bees had been pollinating the North American flowering plants.

Solitary bees (i.e., bees that do not live in colonies and have no social caste system) include some beautiful metallic-colored species. Sometimes large numbers of solitary individuals nest close together and share a common exit hole to their burrows, giving the appearance that these bees are social insects. However, they do not cooperate beyond simple constructing of adjacent burrows.

Bumblebees *(Bombus)* are social bees easily identified by their large, heavy bodies. Bumblebees also nest mainly in the ground, taking over abandoned mouse or bird nests. Bumblebees are some of the most active pollinators in the Rockies. Their colonies are not as large as honey bee colonies. Depending on the species, bumblebee colonies range from 30 to 400 individuals.

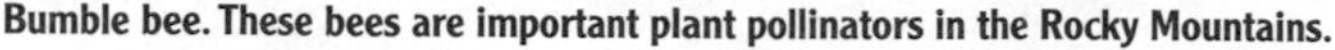

Bumble bee. These bees are important plant pollinators in the Rocky Mountains.

Bumblebees keep their body fluids warm by almost continually vibrating their wings when perched or in their nests, creating the familiar buzzing sound that also warns off predators. Any adaptation that keeps predators at bay is bound to produce imitators in the natural world, and bumblebees have theirs. Flower flies and bee flies are large, hairy flies that often have color patterns similar to bumblebees, and even vibrate their wings to produce the same kind of buzzing sound. These fly families, like all flies, have no sting, but by their appearance and behavior mimic bumblebees so well that most predators (birds and insect-eating mammals) leave them alone.

Moths are also pollinators of plants, but they fly by night, whereas butterflies fly by day. Flowers that attract moths are mostly white, and thus easier to locate in the dark than red, yellow, or blue flowers. Some moth caterpillars are serious pests to coniferous forest trees and can do considerable damage to the flora of the Rocky Mountains.

On the whole, we know far less about the moths of the Rockies than we know about their butterfly cousins. Many species of small, dust-colored moths roaming the hills and valleys have received far less attention from insect collectors than the butterflies, perhaps because many people (including amateur entomologists) consider moths too plain to merit human attention. Unlike the occasional migratory flights of painted-lady butterflies *(Vanessa cardui)* into the Rocky Mountain region, which elicits words of awe and wonder from the human population, the immigration of miller moths (a quite plain-looking moth) into the Rockies elicits cries of anger, grief, and threats of annihilation.

Perhaps the miller moth *(Euxoa auxiliaris)* is a creature only an entomologist could love (and even I am not that fond of them). Actually, the term miller moth is applied to any type of moth that is particularly abundant in and around houses. These moths are called miller moths because the fine scales that easily rub off of their wings reminded people of the dusty flour that would cover the clothing of a miller.

In the southern Rockies, the common miller is the adult form of the army cutworm caterpillar, so-called because, when these caterpillars reach high population numbers, they band together in army-like groups. They may be seen crawling across fields or highways like an invading army.

The army cutworm has an unusual life cycle. The adult moths lay eggs in late summer and early fall in areas of thick herbaceous vegetation. The eggs hatch within a few weeks and the young caterpillars begin to feed. The caterpillars overwinter at low elevations in the Rockies, then resume feeding the following spring, and become full-grown by mid-spring. Then they burrow into the soil and form cocoons. About two to three weeks later, the adult moth emerges. They then migrate and ultimately settle at higher elevations. There they spend one to two months, feeding on nectar and pollinating a number of plants. They rest in sheltered areas, such as under the rocks of talus slopes. In late summer, they return to the lower elevations to lay eggs. The cycle, thus, is renewed.

Biting Flies

Though the Rockies are not plagued by the numbers of biting flies seen in the boreal forest, the arctic, or tropical wetlands, there are places in the Rockies where you may begin to feel persecuted by them, especially near bodies of standing water. Insect repellents are generally effective at keeping the nuisance to a minimum, as is the wearing of long-sleeved shirts and pants (as opposed to shorts and T-shirts). An undeniable fact is that some people attract more biting flies than others. For instance, a host-seeking mosquito is guided to your skin by following the trail of CO_2 gas that comes from your exhaled breath. Once they have landed, they rely on a number of chemical cues to determine if you are an acceptable source for a blood meal. Folic acid is one chemical that appears to be particularly important. Fragrances from hair sprays, perfumes, deodorants, and soap can cover these chemical cues, but they can also enhance your chemical attraction. Dark-colored clothing absorbs heat and makes most people more attractive to mosquitoes. Light colors reflect heat and are generally less attractive. Detergents, fabric softeners, perfumes, and body odor can counteract the effects of color.

Mosquitoes (family Culcidae) are the main group that come to mind when people think of biting flies. Yellowstone National Park has thirty-three known species of mosquitoes, and the Canadian Rockies have at least twenty-eight species of mosquitoes. Mosquito eggs are laid in a variety of kinds of standing water, including ponds, lakes, backwaters, and pools in streams. The larvae hatch and mature over a period of four to six weeks. Mosquito females take blood from birds and mammals in every habitat and elevational zone of the Rockies from the plains to the alpine tundra. Females are the only mosquitoes that bite, using a combination of piercing mouth parts and a suction pump to get blood from the host animal. Blood tends to clot when exposed to air, so mosquitoes inject an anticoagulant chemical into the wound to keep the blood flowing freely. This chemical is most often the irritant that makes mosquito bites itch.

Horse and deer flies (family Tabanidae) do not so gently pierce the skin of their hosts to suck blood; rather, they bite a hole in the flesh of their victims, then suck up the blood. For both kinds of flies, it is again only the females that bite. Males feed solely on the nectar of flowers. The Yellowstone region has thirty-one known species of horse and deer flies. The larvae burrow in soil beneath standing water; it takes one to three years for these to mature into adults.

Biting midges *(Culicoides)* are also called punkies and no-see-ums by Rocky Mountain residents. As these names imply, these flies are very small, usually less than one tenth of 1 inch (3 mm) in length. Despite their diminutive size, these midges deliver a painful, irritating bite. All the life stages of this family seem to dwell very close to water, so the best way to avoid them is to move to higher ground. Many species are small enough to crawl right through standard-mesh mosquito netting, so in biting midge country (mainly the northern Rockies), do not tent-camp too near the water.

Our last biter, black flies (family Simulidae) are the scourge of the boreal forest regions of Canada. These pesky flies can range as far south as northern Montana along the Rocky Mountain chain, although their numbers decline dramatically south of the Lake Louise region of Alberta. Female black flies are universally described in the entomological literature as "voracious biters." They tend to attack the heads of humans, probably because this part of the body is so richly supplied with blood vessels. Their bites swell into painful, itchy welts that can last for days. Fortunately, they only bite in the daytime and rest in the foliage at night.

These biting fly tales should not dissuade anyone from visiting the Rockies. The numbers of flies is usually low, and by taking a few simple precautions, such as using repellents and wearing light-colored clothing, people can often minimize the problem. I cannot say the same of arctic Alaska or the boreal forest of Canada, where these troublesome insects occur in such staggering numbers that it is virtually impossible to avoid them.

Ticks

Rocky Mountain spotted fever: Everyone has heard of it, but few actually get it. In the Rocky Mountains, ticks in the genus *Dermacentor* transmit Rocky Mountain spotted fever (also called spotted tick fever by entomologists).

Spotted fever is caused by an organism called *Rickettsia rickettsii*. Taxonomically, this organism falls somewhere between a virus and a bacterium, making it very unusual. *Rickettsia* replicate inside the cells of their host. Unlike other bacteria, but like viruses, they require a living host (a living cell) to survive. *Rickettsia* from infected mammals live and multiply in the digestive tract of the tick carrier but do not cause disease there. They are transmitted to another mammal, possibly one of another species, by tick bite. Once humans are infected, symptoms begin to appear after three to twelve days. High fever, muscle aches, and cough are followed on about the fourth day by a rash that spreads from the extremities to the trunk and head (hence the name "spotted" fever). If untreated, this disease can progress to more serious symptoms, including delirium, coma, brain damage, and eventual death in about 25 percent of untreated cases. People who have been bitten by a tick and develop tick-fever symptoms should seek medical attention immediately, because of the risk of the infection being spotted fever rather than the milder disease, Colorado tick fever. Fortunately, the spotted fever is carried by only a small percentage of ticks. In the Canadian Rockies, this disease appears to be confined to the eastern slopes (i.e., only in Alberta). Your chances of getting Rocky Mountain spotted fever are slight. The incidence of infection is 8 out of 100,000 people. Treatment by antibiotics usually cures the infection, and complications are rare. The death rate is 5 to 7 percent and is usually a result of delay in seeking treatment. You can minimize your risk of tick bite by wearing hats, long-sleeved shirts, and pants in the mountains and by checking yourself regularly for ticks.

Aquatic Insects

The Rocky Mountains are famous for their trout streams and lakes. Fishing is one of the most important elements of tourism throughout the Rockies, and fly-fishing is becoming even more popular. The hand-tied "fly" with a hook in it is meant to entice the wary trout into thinking it is eating—you guessed it—an insect. Aquatic insects are one of the most important sources of food for trout and other fish in the Rockies. Such groups as mayflies (family Ephemeroptera), caddisflies (family Trichoptera), and stoneflies (family Plectoptera) have larvae that live in the water, and adults frequently hover over the water. Accordingly, some fishing lures imitate the shape, size, and color of the aquatic larvae, and other lures imitate the flying adult stage.

Aquatic insects are significant components of Rocky Mountain lakes and streams. Many aquatic insect larvae feed on detritus (mainly dead plant material), and some shred dead leaves in streams, the first step in the long process of leaf decomposition that releases the nutrients into the water. Other aquatic insects feed on algae, scraping it off of the rocks and twigs that lie below the water line.

If you walk along a typical mountain stream in the Rockies, you will see abundant insect life along the way. The stream courses are natural fly-ways for insects like butterflies and dragonflies. The adults of the mayflies, stoneflies, and caddisflies hover near the water, as their larvae cling to the stones and logs beneath the water. If you look into a quiet pool, you are likely to see the miniature stone cases of caddisfly larvae. These tiny tubes are made of course grains of sand, cemented together by the larva to make a home for itself. You might have to turn over a few stones in the water before you get to see the nymphs (immature stage) of stoneflies and mayflies.

Trout adore these stream insects, in both their aquatic (immature) and flying (adult) forms. These are the models for many of the hand-tied lures used by fly fisherman. As such, these aquatic insects are a vital link in the aquatic food chain. If you are lucky, you might see a trout leap out of the water in an attempt to snatch a caddisfly or mayfly. Of course, the most immediate sensation bombarding your senses as you walk along a mountain stream will be the bubbling, splashing, and cascading of the water over the rocks as it plunges downhill. The ceaseless motion poses a problem for aquatic insects. They face the relentless currents day in and day out, and are eventually washed downstream, in spite of their best efforts. But they have an amazing behavior that helps them keep to one stretch of stream throughout their lives. Each day they crawl *upstream*, against the current, perhaps 165 to 330 feet (50 to 100 m). At night they let themselves drift back downstream, feeding along the way. The nighttime drifting is safer, because the fish generally do not feed then.

Insects in the Cold

What is it that allows some insects to not only survive but to thrive in the cold alpine climate of the high Rockies? Like some Rocky Mountain plants, some insects are able to extend their reproductive cycle beyond a single year. Indeed, some high mountain insects spend several years in the adult stage, reproducing only in years of abundant warmth and good-quality food. Other insect species survive year after year as dormant eggs, at the bottom of tundra ponds or buried under leaf litter or inside the tissues of long-lived tundra plants. When a "good" summer finally comes, the eggs hatch, the immature stages rapidly develop, and the adults mate and lay new eggs to await some future "good."

Many Rocky Mountain insects that live in the tundra produce their own antifreeze, such as glycerol, that allows their body fluids to supercool well below freezing and often as cold as -40°F (-40°C). Other insects, such as some caterpillars, tolerate freezing without any apparent damage. One early naturalist who spent a winter in northern Canada found some caterpillars outside his cabin. They were frozen solid and appeared to be dead. He took them inside, thawed them out by the stove, and they began walking around. He took them back outside, where they froze again, and then repeated the thawing process, doing the caterpillars no apparent harm.

Further research in recent years has shown that tundra insects are not always cold-hardy but that they produce their stores of antifreeze and their tolerance to freezing with the approach of autumn. Some insects, such as certain ground beetles, are intolerant of warm temperatures and die if they are exposed to temperatures above about 50°F (10°C), because the enzymes in their bodies operate best in near-freezing temperatures, and these enzymes cease to function when they get too warm.

Several groups of insects thrive in the alpine tundra environment, including leafhoppers (family Cicadellidae) and grasshoppers (family Acrididae). The high-altitude grasshoppers rely on an extended life span to overcome the rigors of alpine climate. Most low-altitude grasshoppers overwinter in the egg stage. These eggs hatch in spring, then rapidly proceed through the immature stages (nymphs) and become adults within a few weeks. But the maturation process requires warmth, and that is in short supply on the tundra. Lower elevation hoppers become adults in as few as thirty days, but this process can take almost twice as long in the alpine tundra. High-altitude grasshoppers often overwinter as nymphs. This strategy gives them a jumpstart on the maturation process when warm weather arrives in the alpine (June). Some Rocky Mountain species stretch their life cycles into more than one year, passing the first winter in the egg stage and a second winter as a nymph. Adults also occasionally survive the winter. Late-developing species commonly survive until the first severe frost, whereas adults of early developing species may die in mid-summer.

One of the more interesting groups of predators operating at these cold, high elevations are ground beetles in the genus *Nebria*. These beetles live near snowbanks that persist through the summer months, and at night they hunt other insects that have become trapped on the snow surface. One of the main sources of insect prey on snowbanks are moths, butterflies, and flies that live at lower elevations but that are carried aloft by upslope winds, then drop onto mountain tops as the wind carries them up to the Continental Divide region. Some of these unfortunate creatures land on snowbanks, where their bodies are chilled to the point of immobility. All the *Nebria* beetles have to do is to go out and find them. The *Nebria* beetles that live in the alpine zone are physiologically adapted to function at temperatures near freezing, and they keep their bodies raised off of the surface of the snow with their extra-long legs.

Grasshopper Glaciers

Of course, not all Rocky Mountain insects are adapted to live in close proximity to snow and ice, as the *Nebria* meals demonstrate. The story of the migratory locusts in the 1800s is an example worth retelling. Outbreaks of these locusts devastated farms and towns all along the eastern slopes of the Rockies in the years 1874 to 1877. The outbreaks were quite spectacular, with millions of individuals massed in swarms that blackened the skies. The locusts descended on fields to devour every green plant before moving on again. Migrating swarms destroyed crops wherever they landed. One swarm was estimated to contain 124 billion individuals.

In some instances, these locusts swarmed too close to glaciers and large snowbeds in the Rockies. There they perished by the thousands, and their frozen carcasses formed layers in the snow and ice. In a few localities, these locust-filled layers have since melted out of the ice. One such place is Grasshopper Glacier, Montana. Here and in a few other localities in the northern and central Rockies, the frozen remains of century-plus dead migratory locusts can still be seen. This spectacular locust, found in the glaciers, has not been seen alive for many decades, and is thought to be extinct. Geneticists are attempting to study the DNA found in the frozen locusts of the "grasshopper glaciers" as a means of determining their relationships with species still living in the region.

Fish

Fish and fishing draw many visitors to the Rockies every summer, in search of that perfect trout stream or peaceful lake where the clever angler might catch the really big one. The fish of the Rockies remain relatively free of pollutants and therefore safe to eat, making the Rockies a "pristine" fishing hole. Approximately fifty-five fish species are native to parts or all of the Rocky Mountain region, and many more species have been introduced to the Rockies, accidentally and deliberately. Most of the deliberate introductions are game fish, such as brook trout, largemouth bass, and other fish that offer anglers sport. Altogether, nearly thirty species of fish have been introduced into the Rockies during the past hundred years. Some species have been moved from one part of the Rockies to another, such as rainbow trout, which are now found throughout the lakes and streams of the Rockies. Originally they were only native to the western-slope regions of the northern Rockies but were introduced as game fish to other areas.

Native Fishes

Ecologically, the waters of the Rockies can be divided into cold-water and cool-water habitat types as far as fish are concerned. The cold-water fishes of the Rockies prefer water temperatures from 40 to 60°F (5 to 18°C). Although some of these fish can tolerate moderately warm temperatures, their reproduction may be reduced or absent in warm waters. Many of the cold-water species die at temperatures warmer than 75°F (24°C). Nearly all of the cold-water fishes of the Rockies belong to the salmon family. Members of this family include the trout, salmon, and whitefish (all family Salmonidae); arctic grayling *(Thymallus arcticus),* and char *(Salvelinus alpinus).* Most of the salmon that are found in the Rockies are Pacific Ocean fish that come inland to spawn in freshwater streams. Fish that feed in saltwater and breed in freshwater are called anadromous. The salmon species that spawn in the waters of the western slope of the Rockies include the sockeye *(Oncorhynchus nerka),* chinook *(Oncorhynchus tshawytscha),* and coho *(Oncorhynchus kisutch).*

The life cycle of the coho salmon is fairly typical of the anadromous group. The first stage of a coho's life starts about 4 inches (10 cm) below the gravel of a small freshwater stream. The female lays thousands of eggs that incubate over the winter in the gravel, then hatch in early spring (about 120 days after being laid). When the fry emerge from the gravel, they are only 1¼ inches (3 cm) long. The fry remain in their home stream for up to a year, growing to about 4¼ inches (11 cm) long. In the springtime, the young fish swim downstream and enter main streams, and then swim downstream to the ocean. At this stage they are called smolts. They grow rapidly after they reach salt water, but they face increased threats of predation from larger fish in the sea. Smolts tend to migrate northward along the Pacific Coast, reaching coastal Alaskan waters by late summer. The southerly return migration occurs after about one year at sea. The coho salmon then return to their home streams to spawn. Their means of finding their way back to their home stream from the ocean remains somewhat of a mystery, but it is thought that a series of imprints on the young salmon occur when they are initially migrating to the ocean. These imprints probably include water temperature and water chemistry (how the water "tastes" to the fish). By the time they return to freshwater, coho salmon have grown to about 20 inches (50 cm) and weigh an average of about 8½ pounds (4.0 kg). Once in their home stream, the females begin digging out depressions, called nests, in the gravel with their tails. Each female digs three to five nests in which to deposit her eggs. The males seek a mate, then release sperm over the eggs in the nests. When this process is complete, the adult fish live only a few more days. Once dead, their decaying bodies provide nutrients to the stream, helping ensure a resource-rich environment for the fry.

The trout species that are native to the Rockies include the lake *(Salvelinus namaycush),* bull *(Salvelinus confluentus),* cutthroat *(Oncorhynchus clarki),* and rainbow trout *(Oncorhynchus mykiss).* Although lake and rainbow trout are native to the northern Rockies, they have been introduced elsewhere and have come to dominate many waters throughout the Rockies. The bull trout is native to the northern and central Rockies, and the cutthroat is native to all regions of the Rockies. As the only native trout in the southern and central Rockies, cutthroat trout have been the beneficiaries of efforts to bolster their numbers by both state and federal fish and game departments. However, these same agencies continue to stock regional lakes and streams with introduced trout species, in response to the desires of anglers, so the fisheries managers are tasked with two seemingly contradictory mandates: Preserve the cutthroat *and* keep the populations of introduced species high enough to satisfy the anglers.

PETER RISSLER

Cutthroat trout. This is a native species of trout, found in many lakes and streams throughout the Rockies.

There are two native whitefish species, the lake *(Coregonus clupeaformis)* and the mountain whitefish *(Prosopium williamsoni)*. In the Rockies, the lake whitefish is native only to Alberta and British Columbia, although it has also been introduced into lakes in Montana. The mountain whitefish is native to both the central and northern Rockies. The other member of the salmon family that is found in the Rockies is a true northerner with southern populations in the Canadian Rockies, the arctic grayling. This species lives in clear, cold rivers and lakes throughout the boreal and arctic regions of North America west of Hudson Bay. Its range extends down into the Canadian Rockies, and there are also isolated populations in Montana.

Sturgeons are represented by two species in the northern Rockies. The white sturgeon *(Acipenser transmontanus)* is the largest fish in North America. Growing up to 20 feet (6 m) long and weighing up to 1,150 pounds (548 kg), the white sturgeon lives in the estuaries of large rivers and moves upstream to spawn. Because of the building of dams, populations became landlocked in the Columbia River drainage in the twentieth century. The lake sturgeon *(Acipenser fulvescens)* is another giant. It can live up to 100 years, the longest life span of regional fish species. The biggest lake sturgeon reported in Alberta weighed 105 pounds (48 kg) and was 61 inches (155 cm) long. Despite its name, the lake sturgeon is strictly a river fish in the northern Rockies.

The white sturgeon is an anadromous species that is slow-growing. They spawn in the spring and summer and remain in fresh water while young. Older juveniles and adults are commonly found in rivers, estuaries, and the ocean. Anadromous white sturgeon return from the sea to large rivers in the early spring to spawn. Spawning usually takes place in a swift current with a rocky bottom, near rapids. Unlike salmon, white sturgeon can spawn multiple times during their life, and apparently spawn every four to eleven years as they grow and mature. Females produce from 100,000 to several million eggs. The eggs hatch in four days to two weeks, depending on water temperature, and the young fish reach maturity in five to eleven years. Young white sturgeon feed primarily on algae and aquatic insects while remaining in rivers and estuaries. In the ocean, adults feed on fish, crayfish, and shellfish, including clams, amphipods, and shrimp.

There are numerous species in the minnow family that are native to the Rockies, including chubs, dace, shiners, and minnows (all family Cyprinidae). Most of the chub species prefer flowing waters, such as runs and riffles of streams. Dace are more often found in still waters, either lakes and ponds or pools in streams. The minnows also live in quiet waters, such as stream pools, ponds, and lakes. Shiners are found in both pools and runs of streams. These fish feed mainly on aquatic insects and plants, and are eaten by many larger fish, including trout.

We will round out our discussion of Rocky Mountain fish by looking at the perch family. There are three native Rocky Mountain species in this family, the walleye *(Stizostedion vitreum)*, the sauger *(Stizostedion canadense)*, and the yellow perch *(Perca flavescens)*. Darters (family Perdidae), a major group of the perch family, are represented by at least 150 species in North America, yet none live in the Rockies.

The huge mouth and long, sharp canine teeth of the walleye make it a fearsome predator and much sought-after game fish. Walleye live in a wide variety of habitats and thrive in large freshwater lakes and streams. Their large eyes allow them to see well in the low-light conditions of turbid waters. They are commonly found in loose schools associated with the bottom and tend to prefer warm summer water in low-flow areas. Walleye require shallow rocky or gravely bottom areas for spawning and egg development. Young walleye feed on aquatic insects but soon shift to preying on other fishes, including young salmon, bass, perch, stickleback, suckers, whitefish, bullheads, and minnows. The young fish typically grow to more than 3 inches (7.5 cm) in the first year. Males generally reach maturity in two to four years, females in three to six years. Walleye can live ten to twenty years and grow to 25 pounds (12 kg) and lengths greater than 30 inches (75 cm).

Walleye spawning usually occurs in the spring or early summer, when the water temperature is about 40° to 50°F (5° to 10°C). Females commonly spawn in small groups, with more than one male. Spawning usually takes place at night, in shallow water over rocks or gravel substrate. The eggs and sperm are released together into the water column and once fertilized they settle to the bottom. Individual females can produce more than 600,000 eggs. The eggs hatch in twelve to eighteen days, and ten to fifteen days later the young walleye are swimming in the water.

The sauger is a close relative of the walleye, but does not grow quite as large. Its weight seldom exceeds 2 to 4 pounds (1 to 2 kg). Both of these fish are native to the eastern slope of the central and northern Rockies. Like the walleye, the sauger prefer the turbid waters of large, shallow lakes and large, slow-flowing rivers. During the spawning season, it goes up tributary streams or backwaters to spawn. The adults eat other fish, crayfish, other crustaceans, and insects. The young feed on midge larvae and later on mayflies.

During spawning, sauger eggs are deposited at random, fertilized, and left unattended. Incubation takes twelve to eighteen days, depending on water temperature. Young sauger reach a length of about 2 to 4 inches (5 to 10 cm) during the first year, and mature in their third or fourth year of life.

The native Rocky Mountain populations of yellow perch are confined to Alberta and southern British Columbia. The yellow perch is only about half as big as the sauger, but it is sought by anglers because of its tastiness as a pan fish. Yellow perch reach lengths of only 6 to 12 inches (15 to 30 cm) and weigh 6 ounces to 4 pounds (0.2 to 2 kg). Adult females are usually larger than adult males. This species has no canine teeth on the jaws or roof of the mouth. Yellow perch is mostly a lake fish, but it also lives in ponds, slow-moving streams, and rivers. It prefers cool, clear water with summer water temperatures of 65° to 70°F (16° to 20°C). It usually swims at depths less than 30 feet (27 m) but is found in waters as deep as 150 feet (137 m). The larger fish prefer deeper waters, leaving the shallows to the smaller individuals. During different seasons, yellow perch prefer different areas of the lake. In spring, they stay near bottom features, such as rock piles and bottom drop-offs. In the summer, they swim near the edges of submerged plants. In the fall, they congregate near prominent points of land, and in winter, they stay over the flat bottoms.

Yellow perch eat small fishes, including shiners *(Richardsonius)* and minnows, smelt, trout-perch *(Percopsis omiscomaycus),* and even the juveniles of their own species. They also eat aquatic insects (especially midge larvae and mayflies), crayfish, and snails. They feed by sight, so they eat mainly by day, often moving into the shallows during evening to feed on schools of minnows. In turn, yellow perch is an important prey species of many larger fish.

Yellow perch spawn once a year in early spring, shortly after the ice melts and water temperatures reach 45° to 55°F (7° to 12°C). They do not build nests, but the female deposits a long, ribbon-shaped mass of eggs, covered by a thick sheath, which protects the eggs from infection and predation. The egg masses are deposited over sand bars, aquatic plants, fallen branches, or other debris in the water. As the female deposits the eggs, she is followed by two to twenty-five males who fertilize them. Many of these egg masses are eaten by other fishes, washed up on shore, or stranded by low water. Surviving eggs hatch in twelve to twenty-one days, depending on water temperature. There is no parental care of eggs or fry once they hatch. The young perch school close to weedy areas where food is abundant. They reach about 3 inches long (7.5 cm) in their first summer.

The current health of fish populations in the Rockies is generally good, but game species fare better than nongame species in many areas, simply because regional and federal governments work hard to maintain the game species for anglers. There are many regional conservation issues. Most of them concern the preservation of the various kinds of aquatic habitats needed by Rocky Mountain fish species.

Selected References

Armstrong, D. M. 1987. *Rocky Mountain Mammals: A Handbook of Mammals of Rocky Mountain National Park and Vicinity*. Boulder: Colorado Associated University Press.

Brown, F. M. 1957. *Colorado Butterflies*. Denver: Denver Museum of Natural History.

Burger, J. 1996. Yellowstone's Insect Vampires. *Yellowstone Science* 4: 13–19.

Burt, W. H. 1964. *A Field Guide to the Mammals*. 2d ed. Boston: Houghton-Mifflin.

Finley, R. B. 1958. The Woodrats of Colorado: Distribution and Ecology. *University of Kansas Publications of the Museum of Natural History* 10: 213–552.

Herrero, S. 1985. *Bear Attacks: Their Causes and Avoidance*. New York: Winchester Press.

Johnsgard, P. A. 1986. *Birds of the Rocky Mountains, with Particular Reference to National Parks in the Northern Rocky Mountain Region*. Boulder: Colorado Associated University Press.

Koch, E. D., and C. R. Peterson. 1995. *Amphibians and Reptiles of Yellowstone and Grand Teton National Parks*. Salt Lake City: University of Utah Press.

Milne, L., and M. Milne. 1997. *National Audubon Society Field Guide to North American Insects and Spiders*. New York: Alfred A. Knopf.

Minckley, W. L., and J. E. Deacon. 1991. *Battle against Extinction: Native Fish Management in the American West*. Tucson: University of Arizona Press.

National Geographic Society. 1987. *Field Guide to the Birds of North America*. Washington, D.C.: National Geographic Society.

Osthoff, R. 1998. *Fly Fishing the Rocky Mountain Backcountry*. Mechanicsburg, Penn.: Stackpole Books.

Page, L. M., and B. M. Burr. 1991. *A Field Guide to Freshwater Fishes, North America North of Mexico*. Boston: Houghton-Mifflin.

Pijoan, M. 1985. *Game Fish of the Rocky Mountains: A Guide to Identification and Habitat*. Flagstaff, Ariz.: Northland Press.

Russell, A., and A. Bauer. 1993. *The Amphibians and Reptiles of Alberta*. Edmonton: University of Alberta Press.

Sibley, C. G., and B. L. Monroe, Jr. 1990. *Distribution and Taxonomy of Birds of the World*. New Haven, Conn.: Yale University Press.

Smith, H. M. 1995. *Handbook of Lizards: Lizards of the United States and of Canada*. Ithaca, N.Y.: Cornell University Press.

Stebbins, R. C. 1998. *A Field Guide to Western Reptiles and Amphibians*. 2d ed. Boston: Houghton-Mifflin.

Stolzenburg, W. 1997. The Naked Frog. *Nature Conservancy* (September/October): 24–27.

Tilden, J. W., and A. C. Smith. 1986. *A Field Guide to Western Butterflies* (Peterson Field Guide Series, 33). Boston: Houghton-Mifflin.

Trotter, P. C. 1987. *Cutthroat: Native Trout of the West*. Boulder: Colorado Associated University Press.

6

First Peoples

Paleoindians, Ancient Indians, Clovis people: These are the designated names given to the first peoples of North America. These people were probably the direct descendants of nomadic hunters who entered the New World from Siberia by way of the Bering Land Bridge, sometime before 12,000 BP. Until recently, most archaeologists believed that these early nomadic peoples made their way from Alaska to the unglaciated regions south of the continental ice sheets by way of an ice-free corridor that opened up between the Laurentide and Cordilleran Ice Sheets during deglaciation. This corridor was thought to have developed as an ice-free region just east of the Rocky Mountains in Alberta, providing a path for people to enter the unglaciated regions of the lower forty-eight states before the big ice sheets had melted very far back from their maximum extent. However, the most recent geologic evidence indicates that the northern end of the supposed ice-free corridor was still blocked by ice sheets, even as the first Clovis people were becoming established in New Mexico.

The questions of how North America was peopled do not end there. The discovery of a 12,500-year-old site in southern South America casts doubts on many theories: the Monte Verde site, situated near the coast of Chile. Archaeologist Tom Dillehay has recently published the results of more than a decade of work at this site, and his work provides solid evidence that people were in this region about 1,000 years before the first signs of the Clovis culture in North America. Archaeologist James Dixon has argued that the easiest and quickest route from northeast Asia to the Americas was by boat and that the people who camped at the Monte Verde site were part of a maritime culture that had sailed down the western coast of the Americas, moving inland at various places to establish the Clovis culture and the Paleoindian cultures of South America. This issue is far from settled, but these recent studies are breaking important new ground.

Regardless of the early origins of North American peoples, Clovis hunters made their presence known to us through their distinctive projectile points that archaeologists have used to classify their culture. Clovis spearheads were thinned along the base of both sides (fluted), to allow them to be more easily mounted on to shafts of wood. Clovis artifacts have been found at many sites in and around the Rockies. The youngest Clovis artifacts yet found in North America date to about 11,000 BP, based on radiocarbon analysis of charcoal and animal bones from Clovis hunting campsites. Existing for only 500 years, this culture was relatively short-lived. One reason for this is that many Clovis bands appear to have specialized in mammoth hunting. Mammoth bones have been found in nearly every Clovis site throughout western North America, along with bones of smaller animals. Most of the prey animals taken by Clovis peoples became extinct shortly after the time that the traces of Clovis culture disappeared (give or take a few hundred years). This extinction included several species of mammoth, North American camels, ground sloths, Pleistocene horses, saber-toothed cats, North American lion, giant short-faced bear, and giant Pleistocene bison. These extinctions were probably an important force behind cultural changes in Paleoindian bands at the end of the last ice age.

Although most archaeologists consider the Clovis people to have been big game hunters, others believe that Clovis hunters more often simply scavenged mammoth carcasses. Although scavenging may seem a less noble occupation, it would certainly have been less dangerous. Attacking a full-grown mammoth with spears would have been an extremely hazardous endeavor, because a wounded mammoth was surely an angry mammoth.

Many mammoth bones have been found in ponds, lakes, and bogs throughout North America. This observation has led to additional speculations about mammoth hunting by Clovis peoples. Did the hunters ambush mammoths when they went to their favorite watering hole to drink? A mammoth mired in wet mud would likely have been easier to subdue than a mammoth running across a dry plain. Or perhaps old, sick mammoths lingered near the water, unable to remain with the herd as they neared death. Hunters could have dispatched such an animal, or simply waited for it to die, then butchered it near the pond. A third theory, put forward by paleontologist Daniel Fisher of the University of Michigan, is that Clovis hunters killed mastodons and mammoths, then stored large portions (hind quarters, whole legs, ribs, and so forth) in ponds during the winter; the pond ice would thus preserve the meat until spring. Fisher's theory makes sense, because a mastodon or mammoth carcass would yield a huge quantity of meat that could not have been eaten in a brief period. Also, storing the meat could be advantageous to the Clovis hunters in times when mammoths were unavailable. Mammoth and mastodon jerky may have been staples of Clovis hunting camps in winter.

When the Clovis culture ended, it was soon followed by the Folsom culture. During the Folsom period, hunters on the plains and in mountain parkland began to focus on bison hunting (Pleistocene large-horned bison). The Folsom culture is associated with distinctive "fluted" projectile points first described from a site near Folsom, New Mexico. Fluted points (spear heads) are thinned down the middle on both sides of the blade. The fluting on Folsom points extended nearly the full length of the blade. Based on modern experiments, archaeologists have come to appreciate that this type of projectile point is very difficult to make, requiring careful preparation of a piece of stone. Because the point is thinned on both sides during its manufacture, breakage is commonplace. Comparisons of the numbers of broken versus intact Folsom points at sites along the Rio Grande Valley showed a failure rate of about 25 percent in attempts to flute and finish the points. Fluting of the quality seen in Folsom points became a lost art. Points made in the years after the Folsom period are much more crudely made than either Folsom or Clovis points, a seeming reversal in technological progress. The Folsom people also made items from bone that included needles, beads, and disks, and antler tines from deer and elk were used to fashion stone tools.

How did a band of Paleoindians hunt, and how many people were there in a band? These are questions that are difficult to answer, based on our limited knowledge of their culture. Practically the only artifacts preserved from Paleoindian sites in North America are stone tools. We have little direct evidence for how these people made shelters, what clothes they wore, what plants they gathered, and a thousand other details of their daily lives. Some inferences about numbers of people who hunted together have been drawn from the number of bones of prey animals found together at kill sites, although these sites may have been used repeatedly, and radiocarbon dating cannot distinguish between events that took place within a few centuries. For instance, at a site near Casper, Wyoming, archaeologist George Frison estimated that Paleoindian hunters butchered 42,000 pounds (19,000 kg) of usable bison meat sometime about 10,500 years BP. The bison were trapped and killed apparently by a large band of hunters. Given the amount of meat taken, a large band of people were likely fed from this kill, rather than a small, isolated band of only a few families.

More clues about Paleoindian lifestyles come from a river valley in central Alaska, where archaeologists David Yesner, Chuck Holmes, and Kris Crossen have excavated a site called Broken Mammoth. Here Paleoindians camped, fished, and hunted along the banks of the Tanana River, about 11,500 BP. This site also sheds new light on Paleoindian culture. Paleoindians were not necessarily just big-game hunters. At the Broken Mammoth site, bones of ducks, geese, and other waterfowl have been found, as well as the bones of large mammals and salmon. The site was named "Broken Mammoth" because small tools made of mammoth ivory were discovered there.

Based on these recent discoveries and fragmentary evidence pieced together from a suite of sites throughout the New World, many archaeologists have begun to view the Paleoindians as true hunter-gatherers, rather than as just big game hunters. The relative importance of big and small game in Paleoindian diets is very hard to reconstruct from the archaeological evidence, because the artifacts are so dominated by the projectile points presumably used to hunt big game animals. These people simply did not settle down at one site for very long, so they did not leave the sort of refuse piles, human burials, and habitation sites that provide well-rounded evidence of some later cultures. In most cases, they left us only a few elements of their stone tool kit and a lot of unanswered questions.

The Archaic period began about 8,000 BP. Archaic peoples foraged for plant foods in addition to hunting both small and large game animals. Their broader food base allowed them to exploit the canyon and mountain country of the West much more successfully than their big-game-hunting predecessors had done. In addition to projectile points, they made nets and snares to capture small game such as rabbits and birds. They also developed stone tools to dig up edible plant roots, cut plant stems, and chop and grind vegetable foods. Their adaptation to life in western regions was quite successful; they spread throughout western North America, and their numbers began to rise. Several hundred archaeological sites that date to this period have been discovered in western North America. Certain similarities in the shape and size of various implements found throughout the region attest to some form of regular trade connections between bands of people.

During the late Archaic period (about 3,000 years ago), plants began to be cultivated for the first time in the southern Rockies. At first, people did not depend exclusively on crops for food but continued hunting and gathering. Hunting and gathering had been practiced by all previous cultures in human history, and it was more than just a small step to give it up altogether in favor of plant cultivation and farming. At its inception, plant cultivation was likely an extension of the hunter-gatherer lifestyle. People who noticed that certain plants were good, reliable sources of food would make an effort to improve the quantity or quality of that plant species in a given area, then return to the same sites year after year to harvest crops as part of an annual migration pattern. These cultivation activities might have included planting seeds, preparing the soil, removing competitive plants from a patch of desirable plants (the activity we now call "weeding the garden"), and digging small channels in the soil to direct a stream flow to the desired location. An interesting example of this limited style of cultivation has been discovered by archaeologists working in the Canadian Rockies of southwestern Alberta (Wilson et al., 1988). They found tantalizing evidence of deliberate planting of bitterroot by early peoples of this region. The roots of this plant were an important food resource for multiple bands of people, linked by a common language and customs through historic times.

Cultivation becomes *farming* when the people settle in one location and tend plots of land throughout the year, thereby abandoning a nomadic lifestyle. Farming is inherently risky, because each year's crop may fail, depending on the vagaries of weather, insect pests, fungal attacks, and other factors beyond human control. A single hail storm can still destroy an entire crop of wheat in Kansas; a summer without much rain can still shrivel a crop of corn in Iowa; a winter with too much rain can drown a crop of vegetables in California. So farming has always been a risky business; it was even more risky in its infancy.

Except for the southernmost part of the Rockies, we have little knowledge of life in the Archaic period, because hunter-gatherers moved around a good deal, and the remains of their temporary campsites yield little beside stone tools and occasional animal bones. Rockshelters and caves in the Southwest have provided a greater variety of artifacts than other regions, because of the preservation of perishable materials in these dry, protected environments. Cave sites have yielded basketry, sandals made of plant fibers, blankets made of rabbit fur, and twig figurines. Archaeologists have even described an Archaic hunting decoy left in a dry cave in Nevada. The decoy was made of reeds, ingeniously woven together into the shape of a duck.

The hunter-gatherers of the Rockies apparently had to cope with some serious environmental changes during the mid-Holocene, about 5,000 years ago. Archaeologist James Benedict has studied a suite of sites in the southern Rockies and concluded that the people of this region moved up into the mountains on a more regular basis during the mid-Holocene because of lengthy droughts on the adjacent plains. Benedict believes that Archaic peoples had followed a yearly migration that took them up to the mountains in summer and down to the foothills and plains in the winter.

Early Peoples of the Northern Rockies

Paleoindians occupied much of the Northern Rockies region. Some of their camps, such as the Charlie Lake Cave site in British Columbia, were quite far north in the Rockies (Driver et al., 1996). Vermillion Lake, Alberta, a Folsom camp dating to at least 10,400 BP, has yielded projectile points and other stone tools, in addition to the bones of butchered animals, especially bighorn sheep. This site is on the flanks of Mt. Cory in Banff National Park. At James Pass in the upper Red Deer Valley of Alberta, archaeologists have found evidence of a Paleoindian occupation just before 10,000 BP. This site lies at an elevation of 5,500 feet (1,675 m), and the locality may not have been far from the edge of retreating glacial ice. Early peoples clearly made use of the high country in the northern Rockies, at least as travel routes to adjacent valleys. For instance, at Clearwater Pass, Alberta, a Clovis projectile point was found at a site above treeline at 7,640 feet (2,330 m).

Typically, these Paleoindian sites occur in areas with unobstructed views to neighboring valleys and easy access to water and wood. Modern-day campers need to be aware of this and report to the authorities any archaeological discoveries they make, such as locating stone tools. It may be that the discovery you make unlocks a long-held secret. The use of high mountain regions continued throughout prehistoric times in the Rockies. At Boulder Pass in Glacier National Park, Montana, hunters camped near a glacial moraine that dates to about 2,000 years BP and hunted bighorn sheep using game drives. Game drives generally consisted of two low stone walls, converging over a few hundred meters to funnel game animals to an ambush point. The highlands of Many Glacier Valley contain several game drive walls, as well as cairns, piles of rock left as route markers. Brown's Pass also contains evidence of prehistoric hunting camps. In addition to hunting, ancient peoples developed small quarries in exposed seams of workable rock, from which they dug stones to make tools.

During the past 1,000 years, use of the alpine zone apparently tapered off, possibly because of climatic deterioration. Cooler, wetter conditions led to more permanent snow beds and the growth of glaciers at high elevations, keeping both game animals and hunters off the high terrain.

In St. Mary's Valley on the eastern side of Glacier National Park, an extensive prehistoric base camp has been found along the northwest lakeshore. This site was occupied for as much as 8,000 years, apparently during the fall and spring seasons. Fishing tools have been found there, including sinker stones for nets. Bison and bighorn sheep hunting was also carried out there. Many prehistoric peoples used lakeshores for bison hunting. The animals coming to the lake to drink were driven out into the water and killed.

Native peoples used the whole Glacier National Park region throughout prehistory. They were drawn, just as we are, to the high mountains and lakes. Thus far, forty "vision quest" sites have been found in the park. Vision quest sites are high mountain localities, such as East Flattop Mountain and Squaw Mountain in Glacier

National Park, where the males went to fast, pray, and seek visions and spiritual guidance.

Early Peoples of the Central Rockies

The earliest human visitors to the Yellowstone region were part of the Clovis culture. These people are believed to have been hunter-gatherers who moved their camp 50 to 100 times a year in search of game animals. Although they probably used campsites repeatedly through the years, they left little evidence of their presence. Anthropologists have suggested that Paleoindians of the Yellowstone region may have followed certain seasonal patterns of activity in which they lived on the plains to the east and south of the Yellowstone uplands in the spring, gathering tender shoots and greens. The scenario continues into late spring, when berries, roots, and bulbs were gathered. During summer, people may well have entered the mountain regions, where they gathered berries and pine cones. Seeds were removed from the cones and stored as food for winter. In winter and spring, these people probably hunted buffalo and other large game in the lowland regions, and in summer they hunted mainly bighorn sheep.

The hunting of bighorn sheep deserves special attention, because it seems that this was an essential part of the lives of ancient peoples of the Yellowstone region. There are numerous game-drive walls in the Absaroka Mountains. The remains of nets have been found at some of these sites as well. Bighorn sheep become docile after being netted, whereas other large-game animals struggle fiercely to get out of a net, which is why the nets that were found are thought to have been used for sheep. Once the sheep were trapped in the net, hunters could walk up and club the animals. This method of bighorn sheep hunting continued from Paleoindian through Late Prehistoric and Early Historic times. Bighorn sheep remains are the dominant fossil group in many regional cave deposits, including Mummy Cave, near Cody, Wyoming. More than 13,000 bones and bone fragments have been recovered from Mummy Cave, and the majority of these are bighorn sheep bones.

Our imagery of Paleoindians in the Yellowstone region is clearer than our understanding of earlier peoples. By 10,000 years ago, people in the Yellowstone region became separated into two cultural groups: foothills–mountain peoples and plains–basins peoples. The highlands group hunted bighorn sheep, mule deer, and to a lesser extent bison. The lowlands group developed communal bison hunting techniques, using both natural land forms (cliffs, ravines, and so forth) and corrals to capture and kill animals. The two groups developed distinctive styles of projectile points. It appears that the highland groups were relatively isolated, because there was more local development of styles of points in this region. Some archaeologists believe that drought conditions about 8,000 BP forced the lowlands group to retreat back up to the mountains. Whether or not this is so, regional peoples had by this time developed new styles of projectile points.

The obsidian story, worked out by archaeologist Ken Cannon and colleagues, reveals some fascinating details of the lives of ancient Yellowstone peoples. Obsidian is volcanic glass, formed when molten lava cools very rapidly. Cannon has been tracking down this obsidian as it became worked into points, blades, and other tools. Although obsidian glass may be more easily broken than other rock types used for tool making (such as chert and flint), there is nothing sharper than the edge of an obsidian blade. These blades are considerably sharper than the most finely honed surgical steel. There has been a recent resurgence in the use of obsidian blades, in fact. Some modern surgeons prefer obsidian scalpels over steel scalpels. Obsidian tools lend themselves to archaeological study for two reasons. First, the source of the obsidian can be tracked down because each obsidian deposit is chemically unique, so the source locality of artifacts can be found through chemical fingerprinting. Second, through a method called obsidian hydration dating the age of obsidian artifacts can be determined, even if they are not buried in stratigraphic layers that contain charcoal or other organic materials that can be radiocarbon-dated.

Thus far, Cannon and his research team from the National Park Service have collected more than 500 obsidian artifacts in and around the park. The oldest tool known to have been made from Obsidian Cliff glass is a Folsom point, found in the Bridger-Teton National Forest. The point dates between 10,900 and 10,200 BP. Obsidian was apparently highly valued for tool making. Extensive trade networks developed over time, and obsidian from Obsidian Cliffs found its way as far east as Ohio in Late Prehistoric times, where obsidian artifacts were placed in burial mounds of the Mound Builder (Hopewell) culture. Some Paleoindian artifacts found in the park have been traced to obsidian outcrops from sites at Bear Gulch, Idaho, and Teton Pass, Wyoming. As might be expected, early and mid-Holocene-age obsidian artifacts in the park come mostly from Obsidian Cliffs, whereas late Holocene artifacts show more diverse obsidian sources, reflecting the increasing development of regional trade.

Based on numbers of sites and artifacts, regional human populations increased markedly in the late Holocene, with substantial occupations after 4,500 BP. Additional increases in population took place in the last 2,000 years. American folklore would have it that the Native peoples of northwestern Wyoming scarcely ever visited the Yellowstone region, because they were afraid of the geysers, fumeroles, and other geothermal features. Archaeological research during the past few years has shown otherwise.

Early Peoples of the Southern Rockies

Clovis peoples traveled widely throughout the Rocky Mountain region, including the southern Rockies, leaving little but a few stone tools as evidence of their passing. At several sites in the Rocky Mountain region, Clovis artifacts have been found associated with mammoth bones. We know that Clovis hunters visited the Rocky Mountain National Park region at least occasionally, because a Clovis projectile point was found near treeline at the east end of Trail Ridge. Based on extensive surveys of the Continental Divide regions in the park and farther south in the Indian Peaks Wilderness, archaeologist James Benedict (1992a, 1992b) has suggested that these early Paleoindians spent little time in the high country of Colorado. There are Clovis-age artifacts in this region, but nearly all of them have been found either on mountain passes or in valleys leading to passes. Thus it appears that these early hunters crossed over the mountains but spent little time hunting there.

The original Folsom discovery in New Mexico demonstrated that these people also lived in ancient times, because a Folsom point was found embedded in the bone of an extinct Pleistocene bison *(Bison antiquus)*. The earliest Folsom site in northern Colorado is the Lindenmeier site east of Rocky Mountain National Park in the piedmont of the Front Range. At this site, Folsom fire hearths from a camp date back to about 10,900 BP, which is within 200 years of the age of the original Folsom site in New Mexico. In addition to projectile points, the Lindenmeier site also had scrapers, knives, engraved pieces of bone (including pieces that may have been used in games), and bone needles. People at this site had killed and eaten pronghorn antelope, rabbit, fox, wolf, coyote, turtle, and bison.

Frison described the settlement patterns of Paleoindian peoples in Wyoming. Winter bison hunting was apparently a communal activity, and perhaps several bands of hunters joined forces to hunt bison on the plains east of the Rockies. Evidence of post holes at the Hell Gap and Agate Basin sites indicates that winter camps were made there, adjacent to bison kill sites. During winter, large amounts of bison meat could be frozen for later use. These people apparently made at least occasional trips to the Rockies, because the projectile points that they used to hunt bison on the eastern plains were made from stone that has been traced back to the Wyoming's Big Horn Mountains. So, we know that Folsom peoples were hunting on the plains east of the Rockies at that time, but did they hunt in the high country?

By the end of the Pinedale glaciation (about 10,500 to 10,000 BP), there is good evidence that Folsom hunters were making more use of the high country. The Pleistocene megafauna was gone by then, but other game animals were worth going after, including elk and bighorn sheep. Archaeologist Peggy Jodry of the Smithsonian Institution has recently discovered a Folsom campsite called Black Mountain at 10,160 feet (3,096 m), high in the San Juan Mountains of southern Colorado (Jodry et al., 1996). At this site about 10,800 BP, Folsom hunters fashioned stone tools and perhaps hunted bison or other large game animals. This is the highest elevation Folsom site yet found, and it demonstrates that Folsom people hunted in the high country, much as people did until the late nineteenth century.

To kill animals in rugged mountain terrain, prehistoric hunters of the Rockies erected game-drive walls. Elk, deer, and mountain bison were probably all frequent visitors to the alpine, and from Paleoindian times onward people erected game-drives to hunt them. Today, evidence of these game drives remains on many high slopes of the Rockies as piles of rocks forming a line across a ridge or high valley. When they were built, sticks or brush were probably added to the stone walls, providing more camouflage for the hunters laying in wait for the animals. The object was to drive a group of animals into a narrow point, so that they could be killed more easily. In some places, corrals were built at the ends of the game-drive "funnel." In other places, the animals were directed off a cliff at the end of the game-drive walls. The effectiveness of game drives as a hunting technique is attested to by the number of game drives still visible and the obvious work that went into their construction.

The people who lived in the southern Rockies through the Archaic period (8,000 to 1,500 BP) left few traces of themselves for archaeologists to find. They were hunter-gatherers who probably moved camp many times in a year, following the natural rhythms of the land. These rhythms were attuned to the changing of the seasons and the migration patterns of the animals that these people hunted. By the Late

Prehistoric period, some people had started to settle down in villages, especially the Anasazi people of the Colorado Plateau region.

Primitive varieties of corn and squash may have appeared in the American Southwest by late Archaic times, but evidence of these plants (dried remains in caves) did not appear there until about 4,000 BP. Corn (maize) began as a tropical species that was cultivated in Central America and Mexico. Cold-hardy varieties of corn were slowly developed through a northward progression of peoples, but were not developed for the colder Rocky Mountain regions, so farming was simply not an option for Rocky Mountain inhabitants, except on the southwest margin of the southern Rockies.

The development of corn farming marks the transition from Archaic to the first cultural period clearly associated with the Anasazi people, the Basket Maker II culture. It took another 1,000 years for the practice of agriculture to become firmly established. The oldest dated remains of corn comes from soil dated at 3,120 BP, and seeds of cultivated squash first appear in Anasazi sites about 3,000 BP. The Anasazi agricultural revolution began in earnest about 2,500 BP (500 BC), as dated by tree rings from the remains of pit houses and other structures. Agriculture revolutionized lifestyles, because for the first time, people had the luxury of a stable life in a permanent, settled location. With a reasonable harvest of corn and other crops in the fall, winter no longer had to be a time of living on a knife-edge balanced between survival and starvation.

The Basket-Maker III cultural period began about 400 AD. This cultural horizon is marked in regional archaeological sites by the development of ceramics. Anasazi populations increased during this interval, and people moved to valleys and highlands that were most suitable for farming. Beans also were introduced from Mexico into Anasazi culture during this time. These provided an important source of protein that would otherwise be lacking in a diet dominated by corn alone. Together, these two foods provide a well-balanced mix of proteins and carbohydrates. Also during Basketmaker III times, permanent dwellings were built, described by archaeologists as pit houses. The name "pit house" might suggest a subterranean lifestyle, but actually the pit house comprised a floor level excavated just a few feet below the surface, covered by a building rising more than 3 feet (1 m) above ground. The aboveground portion of pit houses was made of a framework of poles, plastered over with mud. The remains of several pit houses that date from this period are preserved at Mesa Verde, in southwestern Colorado.

From 700 to 900 AD, Anasazi culture became more sophisticated, as expressed in improvements in architecture and ceramics, the development of the ceremonial chamber the *kiva*, and the use of cotton in textiles. Pit houses were built in rows and groups as communities grew. By the end of this period, pit house exteriors were finished with masonry rather than mud. This is a logical step in the evolution from more rudimentary pit houses of Basket Maker III times to the free-standing masonry work used in pueblos and cliff dwellings during the later Pueblo cultural periods.

By Pueblo II times (AD 1000 to 1080), trade routes between Anasazi settlements were well-established, and a cultural continuity of the Anasazi peoples developed that persisted until the demise of most communities several hundred years later. It was during this interval that Anasazi numbers grew to their highest levels, and the outposts of the Anasazi realm expanded to the edges of the Colorado Plateau and beyond. Anasazi settlements ranged west to central Utah and the Virgin River, south to the Mogollon Rim in Arizona, east to the Rio Grande Valley in New Mexico, and north to the foothills of the San Juan Mountains of Colorado.

Climatic conditions during Pueblo II times favored agriculture on the Colorado Plateau, and regional farmers began to make better use of the available moisture through the development of field irrigation and water channeling. More productive strains of corn were also developed. Pueblo II times saw the flowering of civilization at Chaco Canyon, described by archaeologists as the "Chaco Phenomenon." This was a prosperous time for the Anasazi, and they began to move out of pit houses and into small pueblos. These were compact masonry buildings plastered over with clay-rich mud (adobe).

The Pueblo III period (AD 1100 to 1300) marked the apex of Anasazi cultural development. This interval saw the building of the cliff dwellings at Mesa Verde, southern Colorado. Overall, there were fewer settlements during this interval, because people clustered in a few large communities. This demographic shift is similar to the "urbanization" of the United States during the twentieth century, but on a much smaller scale. Many of these larger communities were built on canyon ledges and in shallow caves, such as at Mesa Verde.

During Pueblo III times, Anasazi diets increased in variety. In addition to corn, squash, and prickly pear cactus, these people also ate amaranth seed, cotton seed, beans, pepper grass seed, Indian rice grass seeds, and the seeds of several other wild plants. Foraging for edible wild plants undoubtedly increased as cultivated crops failed under the drought conditions (or perhaps insect infestations?) that occurred in the late 1200s, near the end of Pueblo III.

By 1300 AD, Mesa Verde was abandoned. It was rediscovered in the 1880s by local ranchers who explored the ruins and found many items of everyday use still lying about. The Anasazi departure from Mesa Verde was somewhat more gradual, despite the feeling of "quick abandon" one gets when seeing the artifacts left behind. But why did the Anasazi abandon a seemingly flourishing community? Several theories have been put forward to explain the departure, mostly pointing to a change in climate.

Tree-ring data from the Colorado Plateau region (in Arizona, New Mexico, Colorado, and Utah), analyzed by Jeffrey Dean of the Laboratory of Tree-Ring Research in Tucson, indicate that major droughts occurred throughout this region at 1150 AD, and again between 1250 and 1450 AD. These periods of diminished precipitation would have led to poorer crop yields that eventually forced the Anasazi to abandon Mesa Verde. The timing of this drought interval coincides with what climatologists refer to as the Medieval Warm Period in Europe and elsewhere.

In the 250 years before the droughts (900 to 1150 AD), changes in Anasazi culture coincided with the most benign climate of the late Holocene. Water levels in streams were at a late Holocene maximum, precipitation was increasing, and crop yields became more predictable. Climate models indicate that during this period, the elevational zone in which upland dry farming could be done (the type of farming carried out by the Anasazi on the mesa tops at Mesa Verde) expanded dramatically, giving regional farmers roughly twice the arable land that is available today in the Four Corners region. Unfortunately, this advantageous climate was setting up the Anasazi for the devastating droughts that were to come, because as the population grew, so did the people's dependence on good harvests. By about 1150 AD, large populations were taxing the agricultural capacity of the land. Then severe droughts began to strike the Colorado Plateau.

Between 1100 and 1300 AD the Medieval Warm Period began to subside and was replaced by the Little Ice Age, a time of intense cold in the Northern Hemisphere. The onset of the Little Ice Age saw the growth of mountain glaciers in the Rockies and brought colder, drier weather to the Colorado Plateau. Tree-ring records from Mesa Verde indicate sharp declines in precipitation from 1276 to 1299 AD, a twenty-four-year drought. The elevational zone for upland dry farming began to shrink and may have disappeared altogether by 1300 AD. This cold, dry climate persisted in the region until the mid-1800s, when warmer, wetter conditions finally returned. Before the drought was over, Mesa Verde and many of the other Anasazi settlements had been deserted.

The residents of the plains to the east of the Rockies and the mountain-dwelling people did not engage in agriculture. These people remained hunter-gathers until the time of European contact. They tried to persist in their nomadic way of life, even as the trappers, miners, and settlers began arriving by the tens of thousands. By the end of the nineteenth century, all of the regional Native peoples of the Rockies except the Pueblo had been removed from their traditional homelands and placed on reservations by the United States and Canadian governments. These proud peoples had a long history in the Rockies, and by the time of European contact had established tribal regions throughout much of the Rockies and adjacent regions. There was a great diversity of peoples.

The Blackfeet, a confederacy of peoples who speak Algonquian languages, roamed the northern Great Plains between the upper Missouri and Saskatchewan rivers. The three nations of Blackfeet, the Siksika or Blackfoot proper, the Kainah or Bloods, and the Piegan, refer to themselves as "the Bloods." During the 1700s the Blackfeet hunted bison from Saskatchewan to Montana, and by the mid-1800s they dominated much of the northern Rockies. The Blackfeet lived in teepees (also spelled "tipi") and moved camps often, gathering each summer for social and religious ceremonies. The Blackfeet were expert equestrians and buffalo hunters and fierce warriors. They were much feared by other peoples such as the Cree, Sioux, and Crow. The three Blackfeet peoples would join forces when entering wars with neighboring Native peoples.

The Flatheads, also known as the Salish, originated in the Flathead Lake region of northwestern Montana. The name "Flathead" was given to the Salish by other regional peoples because these other peoples had a practice of shaping the heads of their babies into a peak with a cone-shaped wicker headpiece. The Salish people did not do this, so the heads of their children looked flat by comparison. The Salish were not aggressive, but they defended themselves with great bravery against their enemies, the Blackfeet.

The Nez Percé speak a Sahaptian language, as do other peoples from the Pacific Northwest. The Nez Percé formerly occupied a large territory in southeastern Washington, northeastern Oregon, and central Idaho. The name Nez Percé, meaning "pierced nose," was given by early French explorers to this region. These people had a custom of wearing nose pendants, so the name was possibly derived from this custom. However, a second explanation is based on the sign language of the Plains peoples. One of the Plains signs for the Nez Percé was made by passing the index

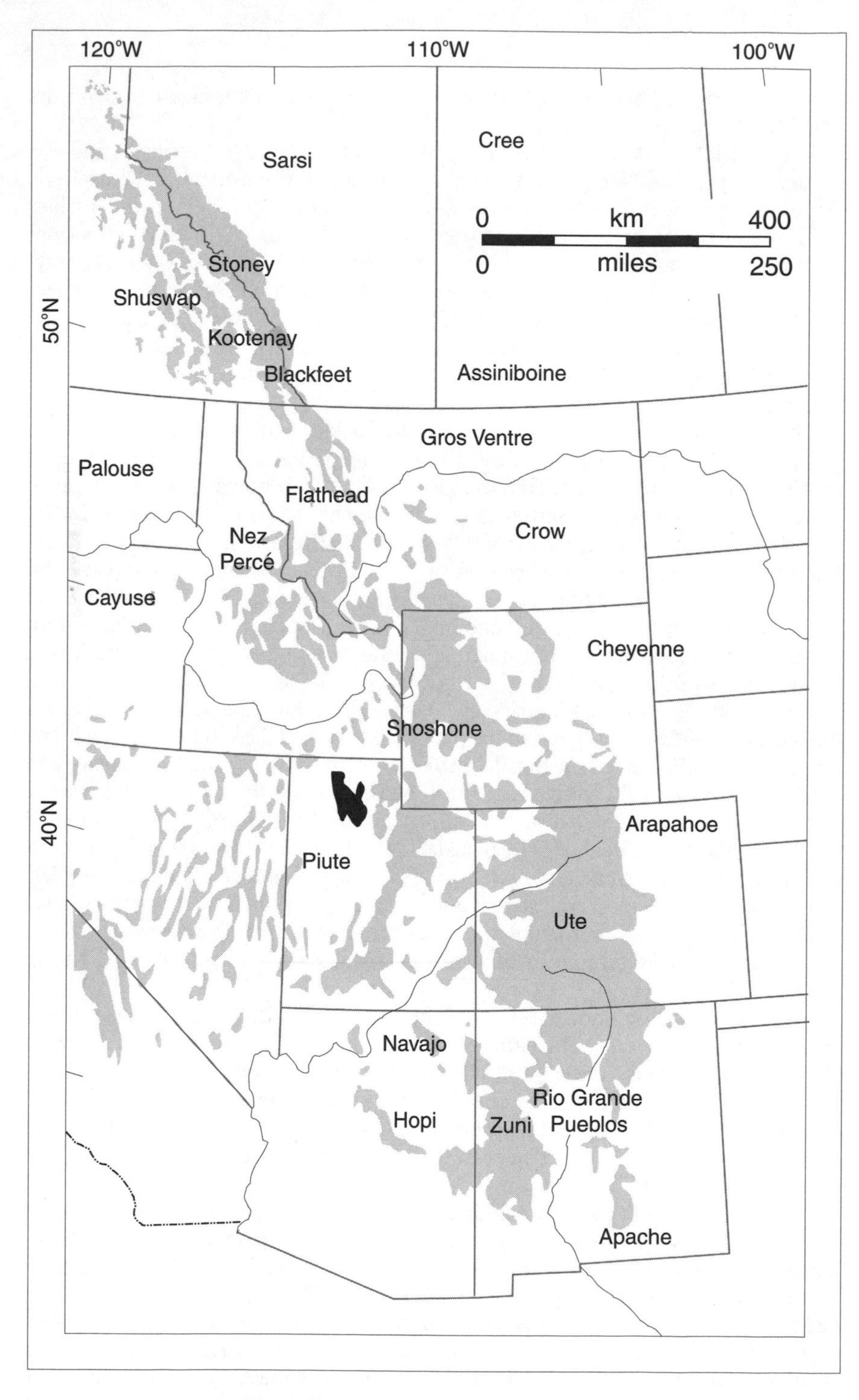

Rocky Mountain region, showing the homelands of Native peoples at the time of first contact with settlers.

finger of the right hand from left to right, close to the nose. This sign probably meant that the Nez Percé warriors would not flinch from an arrow shot right under their noses, but the French Canadians took the sign to mean that these people had pierced noses. This is just one of many examples of nineteenth-century misunderstandings between Native peoples and those of European origin.

Unlike the bison hunters who lived east of the Rockies, the Nez Percé traditionally relied on salmon fishing and on plant foods, such as camas bulbs, wild roots, and berries. However, after about 1700 AD the Nez Percé also had horses and were able to hunt bison more effectively. Their winter camps were built along riverbanks, whereas their summer camps were in the mountains. The Nez Percé had more than forty bands, each led by a popularly selected chief. The Nez Percé had few enemies, but they occasionally battled the Coeur D'Alenes and the Shoshone. Their medicine men treated illness rather than serving as spirit guides. However, the treatment of the sick was a hazardous occupation, because the death of a patient might be followed immediately by the killing of the attending physician by the grieving family.

In response to the Nez Percé's request for instruction in Christianity, a Protestant mission was established at Lapwai, Idaho, in 1837. The Native people were impressed with the missionaries' weapons and wanted to share in the power associated with both the new weapons and the new religion. In 1855 the Nez Percé made a treaty with the United States, ceding the greater portion of their territory. They moved to a reservation that included the Wallowa Valley in Oregon. However, when gold was discovered there, the Nez Percé were forced to surrender all their lands and return to a reservation at Lapwai, Idaho. A band led by Chief Joseph refused to move, and in 1877 he was victorious in a battle with federal troops. Joseph then led his band on a retreat of more than 1,000 miles (1,600 km). Joseph and his band were captured near the Canadian border. They were sent to the Indian Territory (now Oklahoma), where many died. Some of the survivors were finally allowed to return to Idaho, where the majority now lives.

The Shoshone, or Snakes, are a people in the Uto-Aztecan language family that lived in the mountains of western Wyoming and Montana, central and southern Idaho, and parts of Utah, Nevada, and Oregon. The Western Shoshone (also spelled Shoshoni) lived as hunter-gatherers in the Great Basin. This group found its way onto the plains east of the Rockies as early as the 1500s. The Northern Shoshone acquired horses in the mid-1700s and adopted the bison-hunting culture of the Great Plains. Their territory ranged from the upper Snake River to the upper Bighorn River. The famed Sacagawea (1787?–1884) was the Shoshone woman who guided the American explorers Lewis and Clark on their trek over the Rockies. She was probably born in Idaho, but at the age of about 16 to 18 years she was captured by the Hidatsa and sold to a Canadian trapper named Toussaint Charbonneau. He married her in about 1800, and in 1804 Charbonneau was hired as an interpreter and guide for the Lewis and Clark expedition. However, it was Sacajawea who did most of the guiding for the expedition on its trek through western Montana and Idaho, where she saved them from being killed by her people. She died on the Washakie Reservation in Wyoming in 1884.

The Cheyenne speak a language in the Algonquian group. Originally from Minnesota, the Cheyenne were driven out of the area by the Sioux and Ojibwa in the 1600s. The Cheyenne migrated west, eventually ending up in the Black Hills of South Dakota and the plains of Wyoming, where they hunted bison. This move probably took place in the late 1700s, shortly before the Lewis and Clark expedition. Never very numerous, their population was estimated at about 3,500 in 1820. This people's name for itself is Ni-oh-ma-até-a-nin-ya, and the name *Cheyenne* comes from the Sioux word *Shi-hen-na,* meaning "those who paint their faces with red earth." Oral history records suggest that the Cheyenne often fought with the Crows, both before and after settlers arrived in the west. The Cheyenne were considered by most Plains peoples to be fierce warriors, but to the Crows, the Cheyenne were "not brave, just crazy." Some anthropologists believe that none of the Plains peoples were warlike until the arrival of the horse. Horses gave them not only an excellent means of making war but horses themselves became the chief bounty sought from warfare.

Cheyenne hunting capabilities were greatly enhanced by their acquisition of the horse. The acquisition of the horse took place at different times in different parts of the Rocky Mountain region. By about 1830 sufficient numbers of horses had been acquired to give the Cheyenne great mobility. At that time the Cheyenne split up into two main groups: the southern Cheyenne whose territory centered on the upper Arkansas River and the northern Cheyenne whose territory centered on the Platte River. The southern Cheyenne shared the plains of eastern Colorado with the Arapaho. The Colorado gold rush of 1859 brought to the area tens of thousands of immigrant miners who were followed by settlers. After a series of skirmishes between settlers and the Native peoples, U.S. Calvary forces slaughtered whole families of Cheyenne at Sand Creek, Colorado, in 1864. This massacre, of course, outraged the Native peoples.

In 1876 combined forces of Sioux and Cheyenne warriors defeated the troops of General Custer at the Battle of Little Bighorn. It helps to put the battle into perspective by considering the Treaty of Fort Laramie, signed in 1868. This treaty granted the Black Hills area to the Native peoples who traditionally lived there. However, in the early 1870s, rumors of gold in the Black Hills began to circulate in the Eastern United States, and would-be miners began slipping onto the reservation. In 1874, the gold rush began in earnest, and the Sioux and Cheyenne were disgusted with the government's inability to keep the miners from trespassing on their sacred land.

Because the American government did not seem to be keeping the treaty, many of the Native people saw no reason to stay on the reservation. In the winter of 1875 to 1876, thousands of Sioux and Cheyenne left their reservations to go to their traditional hunting grounds in Montana and the Dakotas. The U.S. government warned them that they must all return to the reservations or face military action. Thus the stage was set for the U.S. Army's campaign of the summer of 1876.

The Native peoples joined together to eliminate the U.S. Army from their territories, once and for all. However, they won the battle but lost the war. Knowing that the U.S. Army would inflict severe punishment on the Native peoples for their victory at Little Big Horn, the Native warriors immediately split up and left the region, so that the U.S. Cavalry would have a difficult time finding them. Eventually, they were forced back onto reservations, and their homelands in the Black Hills fell into the hands of the U.S. government. For instance, the Cheyenne surrendered to the U.S. government in 1877 and were sent to Indian Territory.

The Arapaho are part of the Algonquian linguistic stock. The name *Arapaho* is derived from the Pawnee word *Tirapihu,* or "trader." The Arapaho's own name for themselves is *Inuñaina,* or "our people." These people originally inhabited what is now Minnesota, but sometime in the late Prehistoric period they migrated to the plains and eastern foothills of the Rockies. Their territory included the regions between the Yellowstone and Arkansas Rivers. The Arapaho were allied with the Cheyenne, but they were friendlier toward settlers. During the past few centuries, the Cheyenne are thought to have been roughly twice as numerous as the Arapaho. Paintings and early photos of Denver and Boulder, Colorado, show Arapaho teepees in the background, because these locations were part of the traditional camping grounds of the southern Arapaho during the winter. The foothills region was the preferred site for winter camps, because the winter winds are not as strong as they are out on the open plains, and firewood was more available. Southern bands of Arapaho shared the land between the Platte and Arkansas Rivers with the southern Cheyenne. They located their villages on the banks of various streams, especially where clumps of large cottonwoods provided good shade and firewood. The Arapaho had one hated enemy, however: the Utes. The great nineteenth-century Arapaho chief, Left Hand, said that he could not remember a time when his people were not warring against the Utes.

The Arapaho were a part of the Plains culture, which means they often had to move about. As a lightweight, portable shelter, the teepee was essential to this way of life, especially on the prairies, where trees and brush for the construction of other types of shelters were scarce. The Arapaho women raised and packed up the teepees. The framework consisted of lodge poles, beginning with a tripod lashed together at the top, with additional poles added to form a circle. Over this was stretched bison hides, sewn together to make a cover. Most families kept a supply of sixteen to twenty lodgepoles of wood that would not rot, such as cedar.

As bison hunters, the Arapaho were greatly aided by the arrival of the horse in the mid-1700s. Horses escaped or were stolen from the Spanish in Mexico in the 1500s, and by 1600 Native peoples in New Mexico had horses. The plains tribes of Colorado, Kansas, Wyoming, and Nebraska had the horse by about 1720 to 1725. The Crow and Flatheads had the horse by about 1730, and the Cree in Saskatchewan had the horse by about 1750.

The Utes have linguistic ties in the Uto-Aztecan family, linking them with peoples of the Great Basin (such as the Paiutes) and the Aztecs of Mexico. The Ute tribe comprises several bands, including the Tabequache, Muache, Capote, Wiminuche, Yampa, and Uinta, whose traditional homelands include central and western Colorado, eastern Utah, and northwestern New Mexico. As a mountain tribe, the Utes were always hunter-gatherers who had no permanent camps but roamed the southern Rockies on foot, making brush shelters, or wickiups, for nightly camps. Utes had access to a variety of game animals, berries, and other plant foods. Their favorite summer camps were in the central mountain parks of Colorado. Mountain existence was always difficult, and Ute numbers probably never exceeded 10,000 people. They traditionally hunted deer, elk, and bison, but the mainstay of their diet was the jackrabbit, with fish from the lakes and streams in Ute country providing as much as a third of their food. Fishing took place even in the depths of winter. Holes were cut in the ice, and fish were harpooned with spears. Even though the Utes may have had more food resources than the Shoshone, winter existence in any part of the Rockies was difficult for all hunter-gatherers. The Utes relied on roots, nuts, berries, and even mice to get them through the winter.

The Pueblo tribes are thought to be the main descendants of the Anasazi. These peoples live in villages of stone and adobe in northwestern New Mexico and northeastern Arizona and belong to four distinct linguistic groups. However, the cultures of the different villages are closely related. The eastern villages, located along the Rio Grande near Santa Fe and Albuquerque, include Isleta, Jemez, Nambe, Picuris, San Ildefonso, San Juan, Santa Clara, and Taos, whose inhabitants speak Tanoan languages. In the Cochiti, Santa Ana, Santo Domingo, San Felipe, Zia, Acoma, and Laguna pueblos of the Albuquerque region, Keresan languages are spoken. The Hopi live on or near three mesas in northeastern Arizona, and speak a language that is part of the Uto–Aztecan language family. Unlike most Native peoples of the Rocky Mountain region, the Pueblo tribes have remained more or less where they lived when the Spanish arrived in the Rio Grande Valley of New Mexico in 1598. Following in the footsteps of the Anasazi, these peoples have been growing cultivated crops, such as corn, beans, squash, chiles, and cotton, since they arrived in their current settlements, probably in the late 1300s.

By the end of the nineteenth century, nearly all of the Native peoples of North America had either voluntarily surrendered or had been militarily defeated by the governments of the United States or Canada. The lands used by the tribes had consequently been stripped from them. Despite the long history of the Native peoples, it took a mere thirty years, between 1860 and 1890, to bring their lifestyles to an end. Dee Brown, in the book, *Bury My Heart at Wounded Knee*, described that period:

> It was an incredible era of violence, greed, audacity, sentimentality, undirected exuberance, and an almost reverential attitude toward the ideal of personal freedom for those who already had it. During that time the culture and civilization of the American Indian was destroyed. . . .

Sometime about 12,000 years ago, the first humans arrived in the Rockies. The descendants of those cultures remain, despite the attacks of their civilizations, a rich component of the history of the Rockies. Anyone truly wishing to understand the Rockies cannot do so without understanding the peoples who have long called them home.

General References

Bass, A. 1966. *The Arapaho Way: A Memoir of an Indian Boyhood.* New York: Charles M. Potter.

Benedict, J. B. 1992a. Footprints in the Snow: High-Altitude Cultural Ecology of the Colorado Front Range. U.S.A. *Arctic and Alpine Research* 24: 1–16.

———. 1992b. Along the Great Divide: Paleoindian Archaelogy of the High Colorado Front Range. In D. J. Stanford and J. S. Day, eds., *Ice Age Hunters of the Rockies.* Niwot: University Press of Colorado, pp. 343–360.

Brown, D. 1970. *Bury My Heart at Wounded Knee: An Indian History of the American West*. New York: Holt, Rinehart, and Winston.

Coel, M. 1981. *Chief Left Hand, Southern Arapaho.* Norman: University of Oklahoma Press.

Cordell, L. S. 1984. *Prehistory of the Southwest*. New York: Academic Press.

Dixon, E. J. 1993. *Quest for the Origins of the First Americans.* Albuquerque: University of New Mexico Press.

Driver, J. C., M. Handly, K. R. Fladmark, D. E. Nelson, G. M. Sullivan, and R. Preston. 1996. Stratigraphy, Radiocarbon Dating, and Culture History of Charlie Lake Cave, British Columbia. *Arctic* 49: 265–277.

Frison, G. C. 1991. *Prehistoric Hunters of the High Plains.* 2d ed. New York: Academic Press.

Grinnell, G. B. 1972a. *The Cheyenne Indians, Volume 1: History and Society.* Lincoln: University of Nebraska Press. (Reprinted from the Yale University Press edition of 1923)

———. 1972b. *The Cheyenne Indians, Volume 2: Wars, Ceremonies, and Religion.* Lincoln: University of Nebraska Press. (Reprinted from the Yale University Press edition of 1923)

Haines, F. 1955. *The Nez Percés: Tribesmen of the Columbia Plateau.* Norman: University of Oklahoma Press.

Jodry, M. A., M. D. Turner, V. Spero, J. C. Turner, and D. Stanford. 1996. Folsom in the Colorado High Country: The Black Mountain Site. *Current Research in the Pleistocene* 13: 25–27.

Smith, R. D. 1990. *Ouray: Chief of the Utes.* Ouray, Colo.: Wayfinder Press.

Stanford, D. J., and J. S. Day, eds. 1992. *Ice Age Hunters of the Rockies.* Niwot: University Press of Colorado.

Thomas, D. H., J. Miller, R. White, P. Nabokov, and P. J. Deloria. 1993. *The Native Americans: An Illustrated History*. Atlanta: Turner Publishing.

Wilson, M. C., L. V. Hills, B. O. Reeves, and S. A. Aaberg. 1988. Bitterroot *(Lewisia rediviva)* in Southwestern Alberta: Cultural Versus Natural Dispersal. *Canadian Field-Naturalist* 102: 515–522.

7

The Written History

As discussed in Chapter 6, the human history of the Rockies began almost 12,000 years ago. However, because the Native peoples relied on oral histories, the *written* history began only when people of European descent arrived, about 450 years ago. The first European to record his impressions of the Rocky Mountain region was Francisco Vásquez de Coronado. Coronado was a Spanish con quistador who came north from Mexico, searching for the *Cibola*, the so-called seven cities of gold. The region now known as New Mexico was rumored to hold fabulous wealth, and Coronado launched an expedition to find it. Instead, he found Pueblo villages along the Rio Grande Valley. He left this region disappointed, and no new Spanish expeditions were mounted for the next fifty years.

Spanish Colonization of the American Southwest

Half a century after Coronado, Juan de Oñate led a group of Spanish soldiers and priests back into the Rio Grande Valley, establishing a colony there. At first the Pueblo welcomed these newcomers to their land, engaging in trade with them and helping them build their forts and churches. This contact between Spanish and the Native people brought horses to the southern edge of the Rockies, and the Native people soon learned the utility of this "large dog." Within a century, most of the tribes of the Rocky Mountains had horses, a change that transformed their lives. With the introduction of horses, regional tribes enjoyed a sort of golden age, because the horse was a great asset for their nomadic lifestyle.

During the 1600s, Spanish priests forced the Pueblo people to abandon their Native religion and all of the ceremonies they held sacred. The Native people were soon forced to labor in Spanish fields, build Spanish buildings, and follow the Spanish rules, enforced by the sword and the gun. In 1690, resentment resulted in an uprising that forced the Spanish from their strongholds. Battles between the Pueblo tribes and Spanish soldiers, settlers, and missionaries continued off and on for six years. Eventually, Diego de Vargas and an army of 210 men, armed with swords, guns, and cannons, defeated the uprising, and the Pueblos were forced to live under Spanish rule that lasted until New Mexico came under the rule of the United States, a century and a half later.

Realizing their need to develop stronger connections with the Spanish missions in California, two priests, Silvestre Escalante and Francisco Dominguez, set out from Santa Fe in 1776 in search of an overland route to the Pacific. Little did they know that the land between northern New Mexico and California is some of the most rugged terrain of the North American continent. The way west was dissected by endless canyons and blocked by tall mountains. The expedition passed through the San Juan Mountains of southern Colorado, but having gotten as far as Utah Lake, they found that an uncharted desert lay between them and California, so they turned back.

French and English Fur Traders

French explorers began entering the Rocky Mountain region as early as 1724, when Etienne de Bourgomont explored regions west of the Mississippi River. It is thought that he came as far west as the foothills of the Rockies in Colorado. In 1740, Peter and Paul Mallet also crossed the plains to Colorado, then turned south, in hopes of setting up trade connections with the Spanish in Santa Fe. Shortly thereafter, French *voyageurs* (men employed by French Canadian fur companies to transport goods to and from remote stations) followed the Missouri River upstream to southern Montana. Not to be outdone, another fur trading company, the Hudson Bay Company, began to take an interest in the northern Rocky Mountains in 1754. The company sent volunteer Anthony Henday west to what is now Alberta. He recorded seeing the Canadian Rockies from a distance, but he did not reach them. Thus the regions between the Mississippi and the west coast of North America remained essentially unknown to Americans and Canadians before 1800. Meanwhile, to the south, the Spanish were leery of incursions into their New Mexico territory, and rejected the trade offers of the French explorers.

The Pacific Northwest region began to be explored toward the end of the 1700s. Robert Gray, an American, found the mouth of the Columbia River in 1792. Gray was an explorer who was sailing the Pacific Coast when he came to the mouth of the Columbia. Five months later, a British crew under William Broughton sailed 100 miles up the Columbia River and claimed this region for Great Britain. The governments of both the fledgling United States and Britain could easily see that the Columbia watershed was an important land to claim, both for political purposes and because of the wealth of natural resources that it contained. This dispute was not settled until many years later.

By the turn of the nineteenth century, Europeans and Anglo Americans were developing a taste for hats made of beaver fur. This fashion was to launch a whole new episode in the exploration of western North America, as has been discussed previously. Beaver populations in eastern North America were already declining by that time, and fur-trading companies were eager to develop new sources for beaver and other furs. The Hudson Bay Company sent trader Peter Fiddler west of Hudson Bay in search of new fur sources in 1791 to 1792. Fiddler followed the South Saskatchewan River into the Rockies. The big competitor of the Hudson Bay Company in Canada at that time was the North West Company. Not to be outdone, they sent their man, Alexander Mackenzie, across the Canadian Rockies to present-day British Columbia. Mackenzie traveled all the way to the Pacific, taking note of tribes willing to exchange furs for trade goods, such as blankets, iron utensils, rifles, and whisky. By 1797, the Hudson Bay Company had established Ft. St. John, just east of the Rockies on the Peace River; in 1799 the North West Company followed suit with a trading post called Rocky Mountain House, on the North Saskatchewan River. Thus the Rocky Mountain fur trade began in earnest, but in fewer than fifty years most of the streams of western North America had been "trapped out," and the beavers were nearly gone.

The Lewis and Clark Expedition

In 1803 French emperor Napoleon Bonaparte, strapped for cash after his various military campaigns in Europe, sold the remaining French land in North America to the United States. The Louisiana Purchase entailed more than 800,000 square miles (2,100,000 km^2). It included Arkansas, Missouri, Iowa, Minnesota west of the Mississippi, North and South Dakota, Nebraska, Oklahoma, most of Kansas, and parts of Montana, Wyoming, and Colorado. President Thomas Jefferson wasted little time organizing an expedition to take a close look at this newly acquired real estate, especially the Rocky Mountain region. He commissioned Meriwether Lewis and William Clark to lead one of the most famous exploring expeditions ever attempted.

Lewis was 29 years old in 1803. He had served in the army and was a neighbor of Jefferson in Virginia. When Jefferson was elected president, Lewis served as his personal secretary. William Clark was picked to assist Lewis as his partner. Clark was Lewis's friend and was 33 years old in 1803. He had military experience and had taken part in some previous explorations.

Cost overruns are nothing new in government-funded projects. President Jefferson convinced Congress to appropriate $2,500 for the Lewis and Clark expedition. The final cost amounted to $38,000, a lot of money in 1803, but in hindsight, an excellent investment for the knowledge gained.

While still on the East Coast, Lewis accumulated almost two tons of goods using the $2,500 Congress had allocated for the expedition. The following list is only a sampling of the supplies taken west by the Corps of Discovery: Mathematical instruments included a surveyor's compass, hand compass, quadrants, a telescope, thermometers, two sextants, a set of plotting instruments, and a chronometer (needed to calculate longitude). The camping supplies included 150 yards of cloth to be oiled and sewn into tents and sheets, pliers, chisels, thirty steels for striking to make fire, hand saws, hatchets, whetstones, an iron corn mill, two dozen tablespoons, mosquito netting, 10½ pounds of fishing hooks and fishing lines, 12 pounds of soap, 193 pounds of "portable soup" (a thick paste concocted by boiling down beef, eggs, and vegetables), three bushels of salt, writing paper, ink, and crayons. They also brought along gifts to give to the Native peoples along the way, including 12 dozen pocket mirrors, 4,600 sewing needles, 144 small scissors, 10 pounds of sewing thread, silk ribbons, ivory combs, handkerchiefs, yards of bright-colored cloth, 130 rolls of tobacco, tomahawks that doubled as pipes, 288 knives, 8 brass kettles, vermilion face paint, and 33 pounds of tiny beads of assorted colors.

Each member of the expedition was outfitted with clothing, including flannel shirts, coats, frock coats (a man's knee-length coat), shoes, woolen pants, blankets, a knapsack, and stockings. Much of the original clothing simply wore out along the way, and the expedition members resorted to making their own clothing from deer skins and other animal hides, fashioning them in the style of the Native peoples they met along the way.

The expedition went armed with a canon (mounted on the keel boat), fifteen muzzle-loading rifles, knives, 500 rifle flints, 420 pounds of sheet lead for making bullets, 76 pounds of gunpowder packed in fifty-two lead canisters, and an air rifle. This sounds like an enormous amount of ammunition, but in the end they had to ration themselves carefully to avoid running out. The expedition hunted big game as their main source of food along the way, and nearly all of their ammunition was expended in hunting.

The medicines and medical supplies taken on the trip seem rather useless when compared with modern medicine, but it must be remembered that pharmacology was rather primitive in 1803. The expedition carried lots of patent medicines, the formulas of which have mostly been lost by now. Lewis's medical kit included 600 of Dr. Rush's patented "Rush's pills" (effects unknown), a supply of lancets (for bleeding the patient), forceps, syringes, tourniquets, 1,300 doses of physic (purgative), 1,100 doses of emetic (to induce vomiting), 3,500 doses of diaphoretic (sweat inducer), as well as other drugs to produce blistering, increase salivation, and increase kidney output. By the time the expedition was in Montana, Lewis had picked up quite a lot of knowledge from the Native people concerning the use of regional plants for medicinal purposes.

Lewis brought along a traveling library of books he thought might be useful on the expedition. These included Barton's *Elements of Botany*, du Pratz's *History of Louisiana*, Kirwan's *Elements of Mineralogy*, *A Practical Introduction to Spherics and Nautical Astronomy*, *The Nautical Almanac and Astronomical Ephemeris*, a four-volume dictionary, a two-volume edition of Linnaeus (founder of the scientific classification of plants), tables for plotting longitude and latitude, and a map of the Great Bend of the Missouri River.

During the summer of 1803, Lewis supervised the construction of a big keel boat in Pittsburgh. This boat was 55 feet long and 8 feet wide and was capable of carrying 10 tons of supplies. Lewis sailed it down the Ohio River, picking up Clark and some recruits along the way. That fall, the expedition set up camp near St. Louis and remained there through the winter, recruiting and training more men. On May 14, 1804, the expedition set off on their journey with about fifty men, including French Canadians, soldiers, a carpenter, and a tailor. Besides the keel boat, they had two smaller boats called pirogues. The going was not easy as they sailed up Missouri River (against the current). When the wind was unfavorable or calm, they resorted to rowing, poling, and pulling the boats with ropes, walking along the river bank. On May 25 they passed the last village of settlers, La Charette.

On August 3, the expedition had their first official meeting with a Native group. They met delegations of the Oto and Missouri tribes north of present-day Omaha, Nebraska, and handed out peace medals, American flags, and other small gifts, and told the Native people that they now had a "great father" far to the east. They also promised a future of peace and prosperity if the Native peoples would refrain from making war on settlers. Later that month, the expedition held a council with the Sioux, near the location of Yankton, South Dakota.

By early September, the river course took the expedition into Great Plains regions, and they began to see animals unknown in the East, including coyotes, antelope, and mule deer. They killed a prairie dog and prepared it to be shipped back to President Jefferson. In all, the captains described 178 plants and 122 animals that had not previously been recorded for science.

On September 25 the expedition ran into their first trouble with the Native people. Near what is now Pierre, South Dakota, the Lakota Sioux demanded one of the expedition's boats as payment for allowing the explorers to sail farther upriver. This nearly caused a battle, but Chief Black Buffalo calmed his warriors, and no fighting ensued.

By late October, winter cold was setting in, and the expedition made camp across the Missouri River from a large village of the Mandans and Hidatsas. They called their camp Ft. Mandan. It was north of what is now Bismarck, North Dakota. The expedition remained there until April, but in the meantime they hired Toussaint Charbonneau, a French Canadian fur trader living among the Hidatsas, as an interpreter. His young Shoshone wife, Sacagawea, proved invaluable to Lewis and Clark, especially as they entered the Shoshone homelands.

In early April 1804, Lewis and Clark sent the big keel boat and roughly a dozen men back to St. Louis, along with maps, reports, Native artifacts, and boxes of scientific specimens for Jefferson. That same day, the "permanent party" began their journey west, traveling in the two pirogues and six smaller dugout canoes. The expedition totaled thirty-three people, including Charbonneau, Sacagawea, and her baby. Lewis wrote, "We were now about to penetrate a country at least two thousand miles in width, on which the foot of civilized man had never trodden. . . . I could but esteem this moment of my departure as among the most happy of my life."

By late April, the expedition had entered what is now Montana, and they were astonished by the massive herds of buffalo numbering in the tens of thousands. The men subsisted largely on bison meet, each eating 9 pounds of it per day. Near the mouth of the Yellowstone River, Lewis and one of his companions encountered a grizzly bear, never before described for science (see Lewis's description of the event, on p. 92).

By early June, the expedition reached what is now the southern boundary of Glacier National Park, Montana. Here the Missouri River forks, and the captains decided to take the right fork. A few days later, they came on the Great Falls of the Missouri, where they had to portage more than 18 miles (29 km) to get around them all. They made crude carts from cottonwood and buried some of their goods. This difficult portage over rough terrain took them nearly a month to accomplish.

On the 4th of July, the party celebrated Independence Day by dancing late into the night and drinking the last of the whiskey. By the end of the month, the expedition reached the point where the Missouri splits into three forks. By this time, they had entered the outskirts of Shoshone country, and Sacagawea began to recognize some landmarks.

On August 8, Lewis climbed to the crest of the Rockies and stood on the Continental Divide at Lemhi Pass, on the present-day border between Montana and Idaho. He had hoped to see a vast plain to the west, with a large river flowing to the Pacific. Instead, all he saw was more mountains. About a week later, their fortunes improved greatly as they found a Shoshone village. The local chief turned out to be Sacagawea's brother, who allowed Lewis and Clark to trade for horses—a much speedier way to cross the remaining mountains.

At the end of August, the expedition set off with twenty-nine horses, a mule, and a Shoshone guide they called Old Toby. They headed north over a mountain pass and into the Bitterroot Valley. By the middle of September, the Corps of Discovery had climbed into the Bitterroot Mountains, but their guide lost the trail and they began to run short of provisions. Finally, by late September, the half-starved explorers staggered out of the Bitterroot Mountains near modern-day Weippe, Idaho. Fortunately for the expedition, the local tribe, the Nez Percé, befriended them. The men regained their strength, eating salmon and camas roots. Chief Twisted Hair showed them how to use fire to hollow out pine trees and make new canoes.

In early October, the expedition launched five new canoes into the Clearwater River, and in a few days they reached first the Snake and then the Columbia River. After months of frustration and very slow going, they were finally making real progress. By October 18, Clark saw Oregon's Mt. Hood in the distance. This was a landmark named by a British sea captain in 1792, so seeing it reassured the captains that they were at last approaching the Pacific Ocean.

By early November, the expedition was within 20 miles (30 km) of the ocean. At this point, Clark estimated that they had traveled 4,162 miles (6,712 km) from the mouth of the Missouri. His estimate, based on dead reckoning, turned out to be within 40 miles (60 km) of the actual distance.

On November 24 the members of the expedition took a vote and decided to build winter quarters on the south side of the Columbia. They called it Ft. Clatsop, after a local tribe. It was situated near modern-day Astoria, Oregon. On March 23, 1805, the expedition left Ft. Clatsop and started the long trek back to St. Louis. By May they had reached the Nez Percé, but they had to wait for several weeks before crossing the Bitterroot Mountains because of heavy snows. Lewis called the Nez Percé "the most hospitable, honest and sincere people that we have met with in our voyage."

By early July, the expedition had traversed the Bitterroots, and then they split into two groups, so that they could explore more of the Louisiana Territory. Clark led a group down the Yellowstone River, and Lewis and his group took a shortcut to the Great Falls of the Missouri, then they explored the northernmost reaches of the Marias River, not far from the modern United States–Canadian border.

On July 25, Clark's group entered the Great Plains, where they built two dugout canoes. They were just east of present-day Billings, Montana. Clark inscribed his name and the date on a rock face there, the only sign of the Lewis and Clark expedition that remains visible to this day. At the same time, Lewis and three other men were 300 miles north, near the location of Cut Bank, Montana. They got in a skirmish with a group of Blackfeet warriors when they tried to steal some of Lewis's horses, and two Blackfeet were killed. This was the only battle and the only blood shed during the entire expedition. Following this encounter, the explorers rode for twenty-four hours straight, at which point they met the group with the canoes on the Missouri and paddled off to rendezvous with Clark. The expedition was reunited downstream from the mouth of the Yellowstone on August 12.

Two days later, they returned to the Mandan villages. John Colter left the expedition to try his hand at trapping beaver in the Yellowstone region, becoming one of the first American "mountain men." Charbonneau, Sacagawea, and Baptiste remained with the Mandans.

In early September, the remaining members of the Corps of Discovery sped down the Missouri, covering as much as 70 miles (115 km) per day. By September 20, they reached the settlement of La Charette, and on September 20, they reached St. Louis.

Lewis and Clark were treated as national heroes that fall. They were wined and dined in Washington, D.C. Their men received double pay and 320 acres of land as rewards; the captains each got 1,600 acres of land. Lewis was named governor of the Louisiana Territory. Clark was made Indian agent for the West and brigadier general of the territory's militia. Sadly, these events seemed to overwhelm Lewis. Traveling through Tennessee on his way from St. Louis to Washington, Lewis committed suicide at an inn near Nashville. Clark lived another thirty-two years, witnessing the first waves of westward expansion of the United States into the territory he helped to explore for President Jefferson.

The Pike Expedition to the Southern Rockies

Within two years, the first of the free-wheeling fur trappers known as "Mountain Men" was headed west. Colter explored and trapped beaver in the Yellowstone region. The tall tales for which the mountain men were famous began with Colter's descriptions of geysers, mud pots, and petrified trees.

Just after the return of Lewis and Clark to St. Louis, Zebulon Pike led a small group of U.S. Army soldiers to the southern Rockies. He tried and failed to climb one of Colorado's highest mountains (14,110 ft; 4300 m). His failure did not, however, detract from his fame, and now one can "out climb" Pike using an automobile to ascend the famous peak named after him. Soon after his mountaineering attempt, Pike was captured by the Spanish when he wandered too close to their lands south of the Arkansas River. The governor of the Spanish colony sent him home the long way, via Chihuahua, Mexico. Once he got home, Pike proceeded to write his memoirs. Even though his return to the states came well after that of Lewis and Clark, Pike was able to get his memoirs published in 1810, four years ahead of the first edited volume of the Lewis and Clark journals. Thus, Pike's memoir was the first published account of the Rockies in the United States.

Development of the Santa Fe Trail

By 1820, Americans were working out a route from Missouri to Santa Fe, New Mexico. Spanish resistance to American commerce was breaking down in the early 1800s, and the first wagonloads of trade goods began arriving in Santa Fe, along what came to be known as the Santa Fe Trail. Thus began the "Americanization" of New Mexico, a process that would not be fully completed until 1912, when New Mexico was granted statehood. For sixty years, the Santa Fe Trail was one of the most important trade routes in North America, affecting commerce as far away as New York and London. The trail ran 900 miles (1,450 km) across the Great Plains, linking the western edge of the United States at St. Louis, Missouri, and Santa Fe, in what was then part of Mexico. The trail was used not only by traders but also by the U.S. military, along with gold seekers heading for California and Colorado, emigrants, adventurers, mountain men, hunters, guides, packers, translators, invalids, reporters, and Mexican children bound for American schools. This never-ceasing flow of humanity brought great disruption to the lives of Native peoples along the route. Their cultures never recovered from the effects of the trail.

Before the 1800s, Spain jealously protected the borders of its New Mexico colony, prohibiting manufacturing and international trade. This stranglehold on commerce between the United States and New Mexico began to ease in 1819, when a financial panic created a need for hard currency in the Missouri Territory. The Adams–Onis Treaty between the United States and Spain made the Arkansas River (in southern Colorado) the new international boundary. Then in 1821, Mexico won its independence from Spain, and the Mexican government began to look to the United States as a trading partner. That year, William Becknell's party from Missouri was welcomed in Santa Fe. At that time, Santa Fe was an isolated provincial capital, starved for manufactured goods and supplies after years of Spanish trade embargoes. In short order, the Santa Fe trade began to boom, and traders from Missouri sought a reliable route between St. Louis and Santa Fe. Four years later, Sen. Thomas Hart Benton of Missouri persuaded the federal government to survey the trail.

The trail soon became a very popular trade route, and forts began to spring up along the way. These forts offered a safe place for travelers to rest, water their livestock, and do a little trading with the locals. The first of these forts was Bent's Fort, built on the banks of the Arkansas River in what is now southern Colorado in 1833 to 1834. The fort was built by William and Charles Bent and Ceran St. Vrain, entrepreneurs who wanted to cash in on the trade boom brought by the trail.

Westward Expansion of the United States

In 1846, the United States invaded Mexico, and the two countries were at war for two years. The 1848 peace treaty (Treaty of Guadalupe Hidalgo) forced Mexico to give up almost half its lands to the United States, including New Mexico and the other southwestern states. This gave another boost to the Santa Fe Trail, because the whole route was now American territory. Many participants in the 1849 California gold rush used the trail to travel west, and more followed in the 1859 gold rush in Colorado. A second fort was built along the trail in 1851. Ft. Union was built by the U.S. Army to protect people using the trail from the depredations of Apaches and others who raided the wagon trains as they passed through New Mexico.

The railroads eventually spelled the end of the Sante Fe Trail. This process began in 1869, when the trail grew shorter as railroads pushed westward into Kansas. Then in 1878 the railroad reached Raton Pass, the border between Colorado and New Mexico. Two years later the railroad reached Santa Fe, and commerce on the trail came to an end.

The "Mountain Men"

In the early 1800s, the fur trade was beginning to catch hold in the Pacific Northwest, and the time was right for new outposts to be established. By about 1828, the fur trade in the west was flourishing. Fur trappers began meeting at annual trade

fairs, called "The Rendezvous." Two competing fur companies took control of this growth industry, the American Fur Company and the Rocky Mountain Fur Company. In exchange for the trapper's bales of furs, these companies would pay out gold and provide the essentials for another year's trapping: traps, blankets, rifles, powder and shot, flour, sugar, beans, and other necessities. The fur trapper's life was often hard, and the pay often was consumed in a very short time at the annual rendezvous, where the whiskey flowed freely and the parties could last for days on end. In his account of life in the Rocky Mountains, *The Journal of a Trapper*, Osborne Russell wrote about the inequities of the fur-trading business, as the trappers got their yearly shipment of supplies from the fur trading company wagon train, during their annual rendezvous:

> July 5th a party arrived from the States with supplies. The cavalcade consisted of forty-five men and twenty carts drawn by mules. . . . Joy now beamed in every countenance. Some received letters from their friends and relations; some received the public papers and news of the day; others consoled themselves with the idea of getting a blanket, a cotton shirt or a few pints of coffee and sugar to sweeten it just by way of a treat . . . by paying 2,000 per cent on the first cost by way of accommodation.

Russell trapped in the Rockies for nine years, from 1834 to 1843, and knew all of the principal trappers and guides who worked there. In exchange for this life of hard work and brief pleasure, Russell and the other mountain men enjoyed a freedom few people from civilization ever knew. They roamed the mountains and plains as they pleased; trapped beaver, shot a few ermine, foxes, or hares; then moved on to another valley, another stream. Their biggest worry often came in dealing with the Native peoples, but before the hills were crawling with settlers, most tribes were hospitable to these lonely wanderers. Some took Native wives (whether or not they already had wives back east) and lived out their years as adopted members of Native tribes. Some hired themselves out as guides when the settlers starting heading west of the Missouri River in their covered wagons.

The Oregon Trail

About the same time that the Santa Fe Trail was becoming established, Donald Mackenzie of the North West Company worked out a route from the mouth of the Columbia River to southeastern Idaho. One of the chief competitors of the North West Company was John Jacob Astor, purportedly the world's richest man in 1810. He had also made his fortune from fur trading. After Lewis and Clark returned from the Pacific, Astor sent scouts to try to find a better overland route to the mouth of the Columbia River. His party had great difficulties and did not succeed in finding an easier route west, but on their way back, the leader of the expedition, Robert Stuart, discovered South Pass, a 20-mile-wide gap in the Rocky Mountains in what is now Wyoming. This was the one passage through the Rockies that would allow wagons to get through, and it became a key to the founding of the Oregon Trail.

Astor was not interested in helping settlers travel west, and he kept the existence of South Pass a secret. However, the mountain men he hired to get the furs were free spirits who paid little attention to directives from New York City. One of the tough-

est and most resourceful of these trappers was Jedediah Smith. This is a man who allegedly once sewed his own scalp back onto his head after a bear had torn it off. In 1825, Smith found South Pass again, and he made sure everyone knew about it.

Thus the groundwork had been laid for the Oregon Trail by the late 1820s, but the flood of settlers did not start heading west until almost twenty years later. The first emigrants to reach Oregon, traveling in a covered wagon, were Marcus and Narcissa Whitman. They made the trip in 1836. But the big wave of western migration did not start until 1843, when about a thousand pioneers made the journey.

Explorer John Fremont popularized the Oregon Trail, following his travels along this route in 1842 and 1843. The son-in-law of Thomas Hart Benton (who had organized the government survey of the Santa Fe Trail), Fremont was commissioned to write his memoirs for publication in pamphlets and in Eastern newspapers. Benton told him to "make the west seem attractive, and worth settling." Fremont's reports made the Oregon Trail sound like a smooth, level road. Soon the easterners were buying teams of oxen and covered wagons and heading west.

But the journey west on the Oregon Trail was no cakewalk. One in ten of the emigrants died along the way, mostly from cholera, poor sanitation, and accidental gunshots. Many walked barefoot over the entire distance (2,000 mi or 3,200 km). Modern folklore suggests that the wagon trains were constantly attacked by Native peoples, but in reality most Native people actually helped the emigrants.

The 1843 wagon train, called "the great migration," started the mass use of the Oregon Trail. During the twenty-five years that followed, more than 500,000 people used the trail, some of them heading all the way to Oregon but most of them split ting off the trail, heading for California, especially during the gold rush of 1849. As with the Santa Fe Trail, the Oregon Trail quietly fell into disuse when the railway pushed west to the Pacific. The transcontinental railroad was completed in 1869. Today all that remains of the Oregon Trail are patches of ground where wagon ruts can still be seen.

As the settlers started heading west in the 1840s, the era of the mountain man was all but over. The beaver had been essentially wiped out, and men's fashions were shifting from beaver to silk hats. Still, the fur trappers had established new paths into the wilderness. They had built a number of forts that were to have great strategic importance for the next flood of immigrants. These included Bent's Fort, Ft. St. Vrain, and Ft. Lupton in Colorado; and Ft. Laramie and Ft. Bridger in Wyoming. It was not long before these forts became vital waystations and refuges for waves of prospectors and settlers heading for lands west of the Missouri River.

Mormon Settlement of Utah

The first large band of settlers that made their way through South Pass in Wyoming (the lowest, easiest route over the Continental Divide in the Rockies) was a group of Mormons led by Brigham Young. The Mormons had been forcibly driven out of towns east of the Missouri, and decided to strike out for a new land where they would no longer be persecuted for their religious beliefs. The first group arrived in the Wasatch Valley of what would become Utah in 1847. They drove the Utes out of their newly claimed land on the east shores of the Great Salt Lake, and began to plow, plant, and build. As Mormon settlers began to spread out along the foothills of the

Wasatch Mountains, they began to compete with the northern Utes for the scarce resources, such as game, timber, and fresh water. Pushed from the land, the Utes retaliated in a series of raids against Mormon settlements. The Walker War ensued in 1853 to 1854, signaling the beginning of Young's "open hand, mailed fist" policy toward the Northern Utes: Feed the Utes when possible, but fight them when necessary. Between 1855 and 1860, attempts to get the Utes to settle in one place and become farmers failed. Finally in 1861, President Abraham Lincoln set aside two million acres as the Uintah Valley Ute Reservation. The Utes did not go quietly, however. Between 1863 and 1868 they launched a series of raids on settlers, known as the Black Hawk War. Finally in 1869, starving and suffering from Mormon retaliation, the Northern Utes went onto this reservation. Six of the seven Utah bands were relocated there. They are known as the Uintah Band today.

The Gold Rush Era

The Mormon's first few years in Utah were exceedingly hard, but they got an unexpected financial boost by being in the right place at the right time when hordes of gold-hungry prospectors passed through on their way to Sutter's Mill, California, in 1849. The 49ers bought food and supplies from the Mormon colony in Utah. The Mormons would sometimes search the trail heading west across the desert and retrieve some of the trade goods that had been cast off by prospectors lightening their loads. As with all gold booms, there was not enough gold to go around, and many discouraged prospectors left the gold camps of California and headed east again. On their way back through the Colorado Rockies, some of these men discovered gold in Cherry Creek, near its confluence with the South Platte River, in what is now the city of Denver. By 1859 another gold boom was launched.

Soon tens of thousands of men (and a few women) were pouring into the Colorado Rockies, first digging up all the gravel in the streams of the eastern foothills, then moving upstream, finally exhausting the supply of easily obtained placer gold. In the first year alone, some 50,000 prospectors came to Colorado, camping out along the gulches of the Front Range streams, staking claims, and panning for gold. In the next year, gold was discovered in Idaho, and more than 10,000 gold seekers headed to the Idaho gold camps by 1862. In the same year, gold was discovered in Montana, and by 1865 more than 120,000 people hoping to get rich quick had arrived there. Of course most of them found that all the good claims had long since been taken. Also in 1862, gold was discovered on the western slope of the Caribou Mountains in British Columbia. Alas, it was not a major gold strike, but it did mark the beginning of a mining industry there that still flourishes. Indeed, British Columbian mines produced gold worth more than $350 million in 1995 and 1996.

Military Conquest

It is not surprising that the Native peoples of the central and northern Rockies were not pleased when upwards of a quarter of a million prospectors and settlers invaded their tribal homelands. Skirmishes between the Plains tribes and immigrants led to the formation of military units by the U.S. Army that were designed to keep down the Native peoples. One of the first massacres of Native peoples took place at Sand

Creek, southeast of Denver, where in 1864 a U.S. Cavalry unit killed 500 to 600 Cheyenne and Arapaho who were camped near the creek. The massacred Native peoples were by all accounts behaving peaceably when the massacre occurred. The leader of the Army unit, Maj. John Chivington, told the territorial governor that the army had sent them out to kill Indians and that was what they intended to do.

Fighting between Native warriors and the U.S. Cavalry escalated after the Civil War. Troops were sent west to subdue the Native peoples by any means necessary. Native peoples from all around the area were rounded up and shipped off to reservations. Attempts to rebel against these forcible evacuations of homelands were met with sharp reprisals. Finally, in 1876, a group of warriors, mainly Sioux and Cheyenne, attacked Gen. George Custer and his troops at Little Big Horn, South Dakota. For once it was the cavalry that was outnumbered and outgunned. This small victory for the Plains tribes soon turned to disaster, as even more troops were dispatched to punish the Native peoples. The Cheyenne surrendered to the U.S. Army in 1877 and were moved to a reservation in Indian Territory (later to become Oklahoma). In 1879 the Utes were sent from their mountain homelands into a reservation in the desert of eastern Utah. The once-mighty Sioux nation finally laid down their weapons in 1890, when the U.S. Army massacred an encampment at Wounded Knee, in the Dakota Territory.

Advance of the Railroads

One of the greatest agents of change in the westward expansion of the United States was the building of the transcontinental railroad. Working inland from both coasts, the two sets of rails were joined together at Promontory, Utah, in 1869. This link to California from the eastern states, and the spurs that were subsequently built to connect all of the western states to the main line, meant that the months-long journey by Conestoga wagon could now be accomplished in just a few days. Soldiers of the U.S. Army could be rapidly transported to sites where settlers or miners feared clashes with Native peoples. Cargoes of gold, silver, copper, and other precious metals could be shipped east by boxcar. In short, the railroads finally shrank the continent down in size, and the "wild West" ceased being wild shortly thereafter.

The coming of the railroads also drove the diminishing bison herds to the brink of extinction. Hunters working along the rail lines shot and skinned hundreds of thousands of bison on the Great Plains, and the once-mighty herds thundered no more. By 1870 the bison populations had shrunk into two distinct herds, one centered in the northern plains and one on the southern plains. Between 1872 and 1874, hunters shot more than three million bison, shipping 1.4 million hides to markets in the east and wasting the rest. This was a level of wildlife slaughter seldom seen before. W. T. Hornaday (1854–1937), an American zoologist and naturalist, was the director of the New York Zoological Park (now the Bronx Zoo) and a champion of game preservation and laws for the protection of wildlife. In 1887 Hornaday described the causes of the near-extinction of the bison. He felt that the descent of civilization with all of its destructiveness was one of the main causes. Also, he decried humankind's reckless greed, wanton destructiveness, and improvidence in not husbanding such a great natural resource. Hornaday found the absence of protective measures by the federal government to be utterly inexcusable, but he also

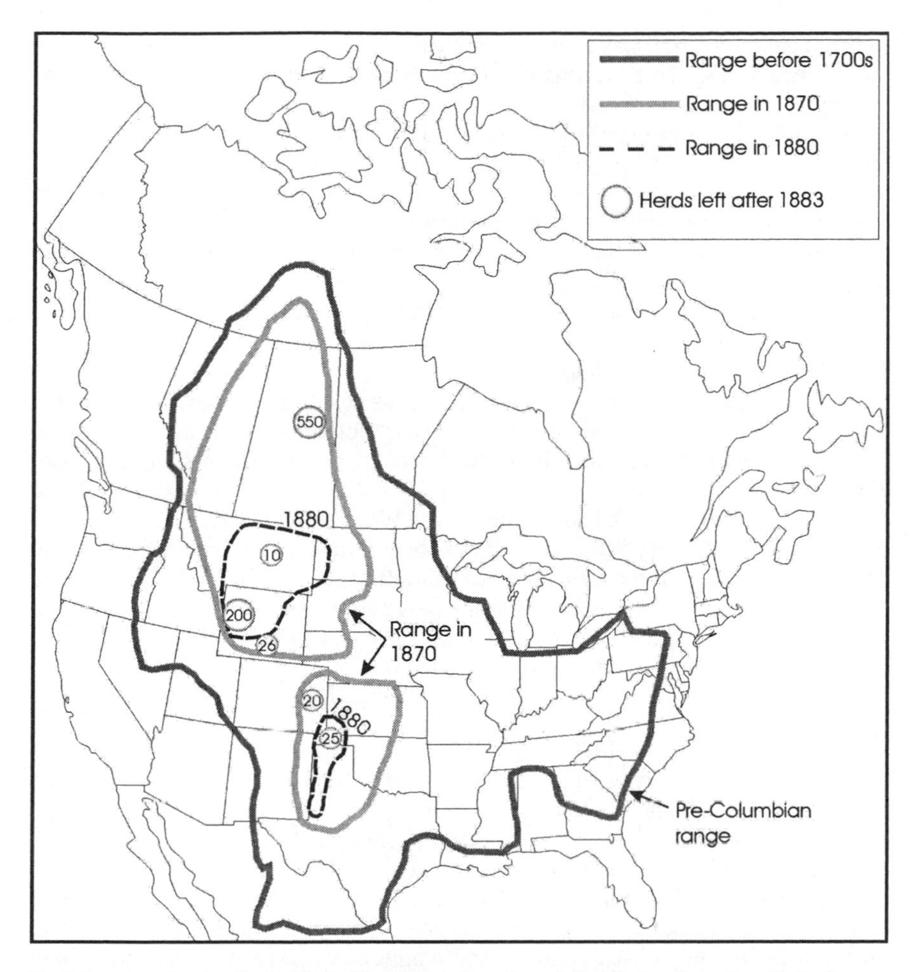

North America, showing the decline in bison populations through the nineteenth century. Illustration after W. T. Hornaday. 1887. The Extermination of the American Bison with a Sketch of Its Recovery and Life History. ***Annual Report of the Board of Regents of the Smithsonian Institution for the Year Ending June 30, 1887:*** **367–548.**

noted that the stupidity of the bison and their lack of fear of humans contributed to their easy slaughter. Finally, the rapid destruction of the bison herds was facilitated by technological innovation. The perfection of the breech-loading rifle and other fire arms allowed one hunter to dispatch thousands of animals per day.

By 1880, only a few small herds remained; these lived in the most remote regions with the most difficult terrain. A herd in the Oklahoma panhandle was estimated to have twenty-five individuals; a herd in eastern Colorado had twenty; two herds in Wyoming had twenty-six and 200 individuals, respectively; a herd in Montana had ten individuals, and a herd in central Alberta had 550. Thus the wild bison population remaining in 1880 was about 831 individuals. Of course this loss of a primary food source devastated the remaining Plains tribes and left them literally begging for food.

In place of the unbroken seas of grass and herds of bison on the plains, farmers and ranchers brought the plough, domesticated cattle and sheep, and a gritty determination to make a living on the land. Water was a critical asset in both farming and ranching, and communities near the Rockies began building canals and ditches to carry mountain runoff to the parched plains below. During the 1870s, towns began to spring up along the Rockies that had little or nothing to do with any gold rush. The lust for gold had not diminished, and indeed, mining for precious metals continues to this day. Yet, as the 1800s came to an end, more people began coming west with other occupations in mind.

Another type of metal, not quite as precious as gold, made news in 1881. Unbelievably rich deposits of copper were discovered in Butte, Montana, and the mines of Butte produced more than six million pounds of copper in their first decade of operation. Montana was granted statehood in 1889, followed the next year by Wyoming and Idaho. These were still boom times in the West, and the settlers and their descendants in these newly founded states were brimming with optimism and regional pride. The Rocky Mountain states and Canadian Provinces, one by one, became more formal parts of the nations that ruled them. New Mexico, still a relatively quiet backwater territory, was rather late in becoming a state, but in 1912, 314 years after the Spanish founded their colony there, it received the formal status of statehood.

The Rise of Tourism

By the turn of the century, the Native peoples and their beloved buffalo had been cleared off the land, and the Rocky Mountain region, that previously wild and forbidding land, had been woven into the fabric of Western civilization. The great railroad companies began laying tracks to the newly created national parks, such as Glacier (founded 1910), Yellowstone (founded 1872), and Grand Canyon (founded 1908) National Parks in the United States and Banff National Park in Alberta (founded 1885). The railroad companies also built luxurious hotels in or near these parks, and the tourists started to come in ever-larger numbers. The advent of automobiles brought even more tourists, and soon the mountain states and provinces began to look on tourism as a vital part of their economies.

The combined populations of the Rocky Mountain states and provinces totaled roughly 1.6 million people in 1900. Populations in most parts of the Rockies rose steadily until World War II, reaching a combined total of about 4.5 million by 1940. In the postwar years, populations began booming, especially in Colorado, Alberta, and British Columbia. Colorado's population went from about 1.2 million in 1940 to 4.3 million in 2000. British Columbia's population rose from about 800,000 in 1940 to 4 million in 2000, and Alberta's population rose from 800,000 in 1940 to about 3 million in 2000. The population of the Rockies has climbed steadily in the past fifty years and now totals about 18 million. This is a tenfold increase in human population of the Rockies in the past hundred years. This rate of growth greatly exceeds the national averages for both the United States and Canada.

In Summary

The written history of the Rocky Mountain region had a somewhat tumultuous start, with larger-than-life mountain men, gold rushes, and Native wars. The railroads and the settlers they brought ushered in a tamer way of life, with the Rocky Mountain region remaining mostly on the back burner of events in North America for the first half of the twentieth century. Now this mountainous region has been "discovered" by the people of both the Atlantic and Pacific Coasts and is experiencing unprecedented population growth.

Selected References

Abbott, C. 1994. *Colorado: A History of the Centennial State*. Niwot: University Press of Colorado.
Ambrose, S. E. 1997. *Undaunted Courage: Meriwether Lewis, Thomas Jefferson, and the Opening of the American West.* New York: Simon and Schuster.
Arrington, L. J. 1994. *History of Idaho* (2 vols.). Moscow: University of Idaho Press.
Barman, J. 1991. *The West beyond the West: A History of British Columbia.* Toronto: University of Toronto Press.
DeVoto, B., ed. 1981. *The Journals of Lewis and Clark*. Boston: Houghton-Mifflin.
Jenkins, M. E., and A. H. Schroeder. 1974. *A Brief History of New Mexico*. Albuquerque: University of New Mexico Press.
Lavender, D. 1975. *The Rockies.* New York: Harper and Row.
Russell, O. 1965. *Journal of a Trapper, 1834–1843* (A. L. Haines, ed.). Lincoln: University of Nebraska Press.
Spence, C. C. 1978. *Montana: A Bicentennial History.* New York: Norton.
Sodaro, C., and R. Adams. 1986. *Frontier Spirit: The Story of Wyoming.* Boulder, CO: Johnson Books.

Epilogue

A Legacy to Preserve

The Rockies represent a chain of habitat islands of coniferous forest, often topped with tundra, and surrounded by a "sea" of desert and prairie. Seeing the mountain chain as habitat islands similar to oceanic islands is a prerequisite to preserving the wonder of the Rockies for future generations.

Today, no one would dream of allowing the kind of ecological disaster that occurred in many parts of the Rockies in the 1800s. Creek beds were turned inside out in search of gold nuggets. Whole regions were pockmarked with mines and mine tailings. Forests were cut to make cabins, mine tunnels, railroad ties, and heat buildings through nine months of the year. Deer and elk populations plummeted as commercial hunters shot them to feed hungry mining camps. At that time, no one particularly cared about all of this. After all, they did not intend to *stay* in the Rockies. They just wanted to make a fortune and take it back to civilization. The devastation brought about by mining did not wane until early in the twentieth century, when increased production costs and decreased value of gold and silver forced the closure of most mines. What is the legacy of all this? Today there are more than half a million abandoned mines in the western United States. These mines have produced some seventy billion tons of tailings. Twelve thousand miles (19,000 km) of waterways downstream from these mines have been polluted. The federal government is spending millions of dollars to clean up some of the worst of these poisonous piles, now designated as EPA Superfund sites.

In the Colorado Front Range region, montane forests that were heavily affected by miners during the 1800s have yet to recover to their primeval condition, and probably never will attain that condition because of the number of houses built in these regions during the past thirty years. Because so many people now live there year-round, fire prevention has become a necessity, even though periodic fires are good for the health of the forests. Although many people may agree about this ecological reality, no one wants to live on 5 acres of land with charred, dead trees.

We have learned a lot about the ecology of the Rockies during the past hundred years, but even though the ravages of the gold rushes have abated, we are left with some serious environmental problems. Real estate developers and politicians in the Rocky Mountain region, as elsewhere, like to talk about "sustainable growth." Population biologists would argue that this term is an oxymoron, because if the population growth of any species continues indefinitely, the resources needed to sustain it will become increasingly scarce, and populations will eventually crash. Granted that humans are more clever than other species, can we really say that we will be able to negate this biological principal? If we could succeed in this never-ending growth, what would happen to the Rocky Mountain wilderness? There are ways to safeguard the remaining Rocky Mountain wilderness, but they require the goodwill and hard work of people who care. I hope that this book has provided the reader with some small spark of interest in this beautiful, fascinating region. The flora and fauna of these mountains are tough in some ways but vulnerable in others. Let us give them as much natural, undisturbed space as possible, so that they may continue to flourish for many generations to come.

Index

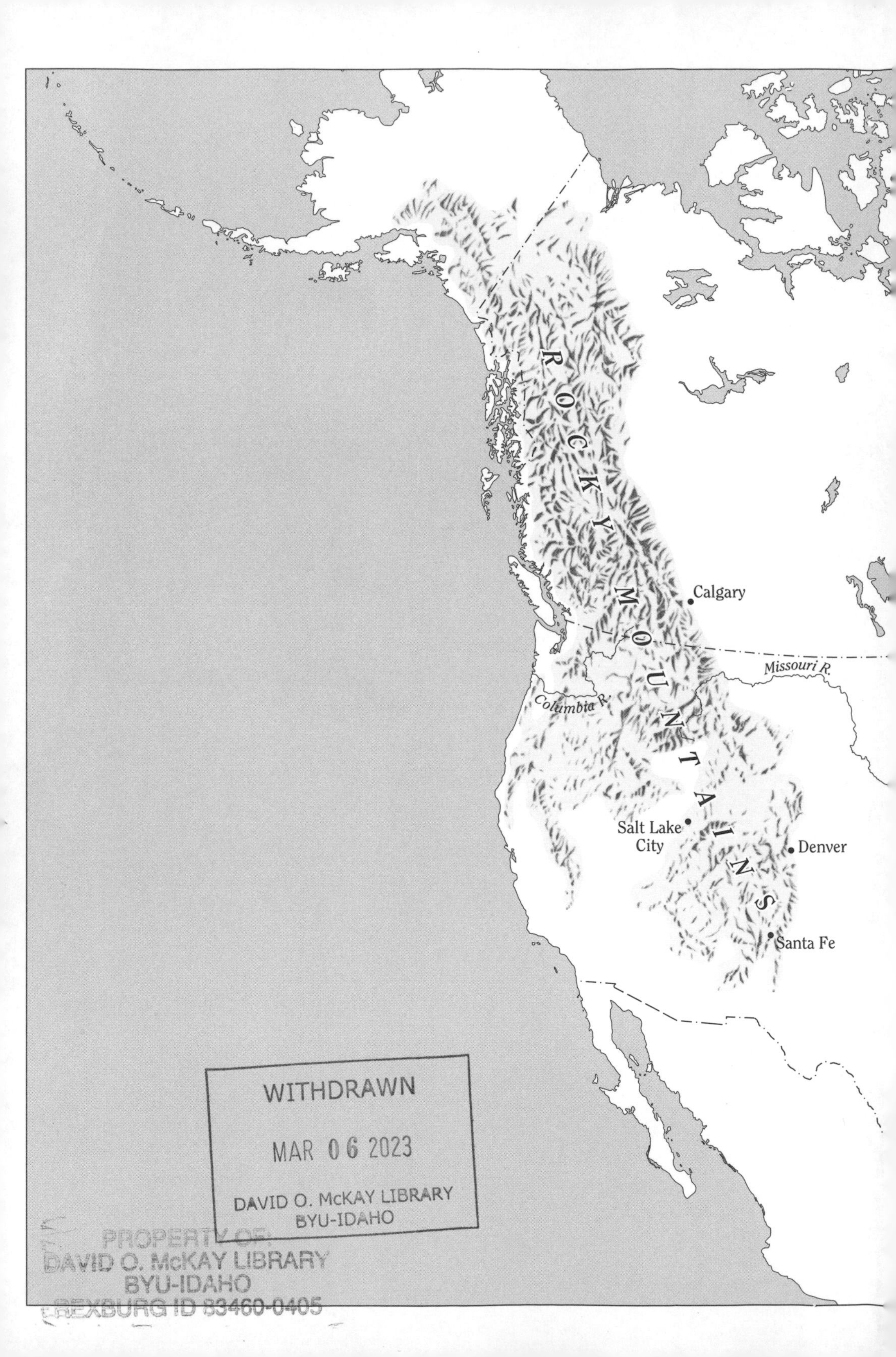
ROCKY MOUNTAINS
Calgary
Missouri R.
Columbia R.
Salt Lake City
Denver
Santa Fe